Nature's Connections

An Exploration of Natural History

Nicola McGirr

Published by The Natural History Museum, London

To David, Caitlin and Gilbert, for the year you gave me to write this book

First published by The Natural History Museum,
Cromwell Road, London SW7 5BD

ISBN 0-565-09144-1

A catalogue record for this book is available from the British Library

Edited by Jonathan Elphick
Designed by Linda Males MSCD
Picture Research by Emily Hedges
Reproduction by Chroma Graphics, Singapore
Printed and Bound by Printer Trento srl, Italy

Cover photograph of false staghorn fern reflected in raindrops, by Whit Bronaugh

Contents

History

Curious Minds

'Sweet are the uses of adversity,
Which, like the toad, ugly and venomous,
Wears yet a precious jewel in his head...'

Act 2, Scene 1, *As You Like it*
William Shakespeare

Fungus gnat in Baltic amber.

Few of us have a specialist scientific background, yet we all have something of the scientist at heart. You only have to watch a baby explore his or her surroundings or listen to a toddler asking endless questions to realize that we are all curious about the world we live in. Curiosity draws millions of people to natural history museums around the world every year. This curiosity may be inspired by current events, prompting questions such as: How can we help to protect endangered species? What is acid rain? Should scientists be allowed to clone organisms or their organs? Or it might come from a love of wildlife programmes, or gardening, or a fascination with the past. You don't have to be an expert to enjoy learning that birds are living descendants of dinosaurs, that the Earth's continents are moving, or that amber has been used for adornment since before 10,000 years BC.

Curiosity is a recurring theme throughout the story of natural history, but it doesn't in itself explain how our interest in the Earth and the life it supports became a science. In a

Greenland whale breaching by Sir William Jardine, 1836. Whales continue to be threatened by human activity.

whistle-stop tour through time, this chapter traces the roots of natural history in Western society from its origins over two and a half million years ago to the end of the 18th century. This survey won't be exhaustive, but it will illustrate how economic and cultural factors have influenced our approach to this subject and our understanding of the diversity of life on an ever-changing Earth.

Before we start our journey, it's worth taking a moment to think about the people who have contributed to the development of natural history.

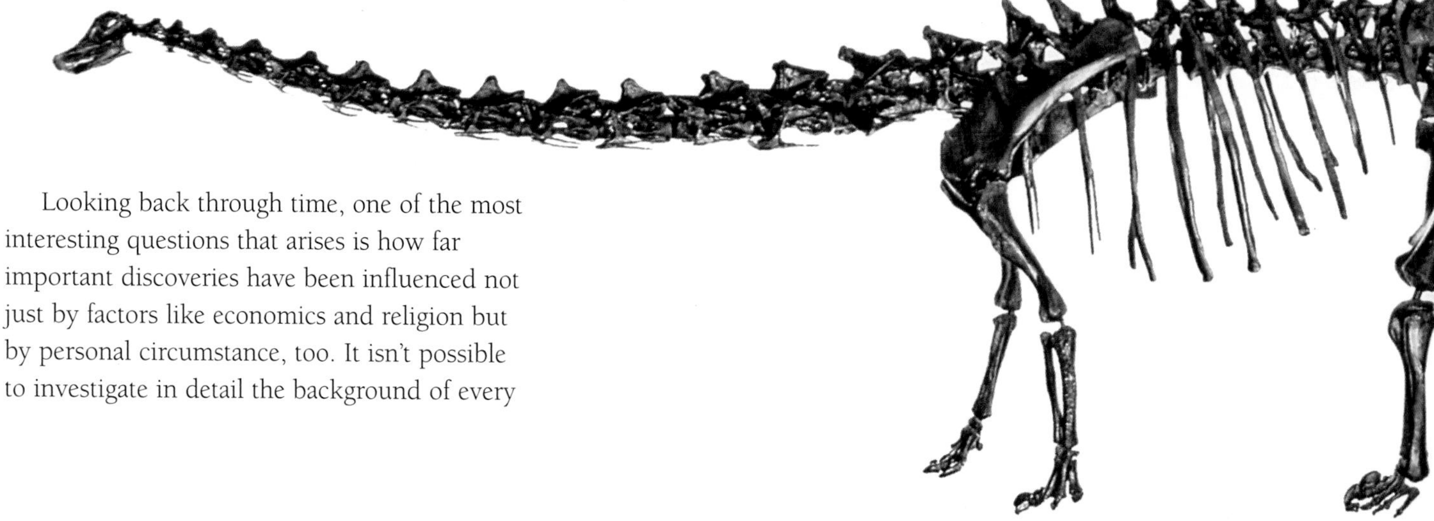

Looking back through time, one of the most interesting questions that arises is how far important discoveries have been influenced not just by factors like economics and religion but by personal circumstance, too. It isn't possible to investigate in detail the background of every naturalist mentioned in this short history, but we will look into the life of Carl Linnaeus, the Swedish naturalist famous for his contribution to the modern system of taxonomy (the science of how living things are classified). The story of Linnaeus gives a fascinating insight into the study of natural history during the 18th century, and provides an example of how it is often an oversimplification to attribute major discoveries to a single individual.

Swedish naturalist Carl Linnaeus, by Magnus Hallman.

First Awakenings 2,400,000 – 599 BC

The roots of modern science, our curiosity with the natural world, can be traced back to our earliest hunter-gatherer ancestors almost two and a half million years ago. Their interest in nature was shaped by the need to solve immediate practical problems, such as how to feed, protect, clothe and heal themselves — a trend that would continue until the more abstract approach of the Ancient Greeks started to supersede it from about the 6th century BC. Understanding plants and their different properties, for example, was vital for success and survival. Which plants were edible, which harmful, and which could be used for practical purposes, like healing the sick or making dyes? The two most ancient natural history traditions, herbal medicine and taxonomy (the science of plant and animal classification) owe their beginnings to these most fundamental of questions.

Hunting and living in groups improved the ability of early humans to provide food and protect one another, and led to a gradual transition from a nomadic lifestyle to the earliest Neolithic village settlements. The archaeological evidence is sketchy, but we can trace the origins of language to this early period in human history, which experts believe was motivated by the need to plan and execute co-ordinated hunts. The development of language was a groundbreaking step in the story of natural history because it offered the means by which new observations, methods and techniques could be taught, learnt and passed on to benefit future generations. Cave paintings and Australian Aboriginal 'X-ray' art provide the earliest record of anatomical observation in images of complete and butchered animals that depict entrails, organs and bones.

By about 8000 years BC, the combined effect of the last ice age and highly efficient hunting techniques had depleted animal reserves and led to a crisis in the availability of food. This crisis encouraged hunter-gatherers to look for alternative food supplies. Years of knowledge, accumulated by observing the behaviour of plants and animals, led to the development of the two fundamental techniques that paved the way for the agricultural revolution: the cultivation of seeds from wild grasses and the domestication of wildlife.

A towering skeleton of *Diplodocus*, reminder of early life.

A recent example of Aboriginal 'X-ray' art. This fish, a barramundi, was painted in the early 20th century in Deaf Adder Gorge, Kakadu National Park, Australia.

These new agricultural techniques gave our ancestors greater control over natural resources, and made it possible to generate enough food to feed large populations from small areas of land. Village communities flourished in fertile valley plains such as the banks of the Nile in Egypt, cleared of forests and with a ready supply of water. Gradually, these villages expanded and united to create the earliest cities.

Cities established a new form of social structure. Surplus food created goods for barter and thriving economies, founded on the forced labour of slaves, allowed people to specialize in areas other than farming. Craftsmen exchanged food and raw materials for finished goods, such as cooking and farming utensils. For the first time, a proportion of the population, freed of the need to create food, could fully occupy new roles within cities as administrators, priests, craftsmen, traders and moneylenders.

Evidence of our fascination with minerals predates the growth of cities. Necklaces and other jewellery made from pieces of metal ore have been found dating from the Stone Age, but it was the growth of trading civilizations like the Egyptians that stimulated the demand for a more systematic understanding of the properties and uses of minerals. The word 'metal' comes from the Greek word meaning 'to search', because only gold and copper exist in their raw state and they are extremely scarce and difficult to find. It was the discovery of the technique for creating alloys that led to the production of bronze, a mixture of copper and tin ore. Our earliest understanding of mineralogy can be traced to the Bronze Age demand for metal ores and the prospecting and mining skills developed by the traders who supplied them.

The trading of small amounts of precious metal created the need for precise measurement

Lehuta torc (necklace), Bronze Age, Hungary.

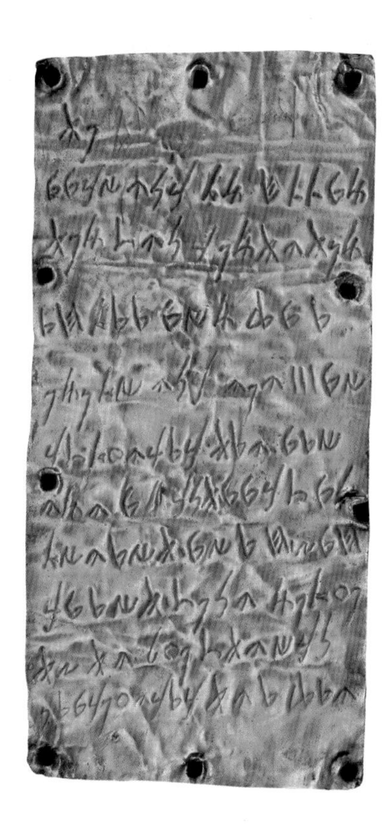

Gold plate with Phoenician inscription, from Santa Severa, Pyrgi, 5th century BC.

and led to the development of weighing scales. Writing emerged in direct response to the need to record these measurements — 'not so much a deliberate invention as an incidental by-product of a strong sense of private property'. The Egyptians were the first to write down their observations, using hieroglyphic symbols, but it was the Phoenicians who created the alphabet, a system that reproduced the sounds of words and created a common method of communication between traders who spoke different languages. For the first time observations could be recorded for future generations in words and diagrams, a vital step towards the accumulation of knowledge upon which every branch of science is based.

Some of the earliest written records, of medical diagnosis and techniques, were made by Egyptian priests. They described setting broken bones and the antiseptic effect of mouldy bread on wounds — an early application of penicillin. Recommended medications included castor oil purgatives, opium poppy seeds for pain relief, and garlic, a powerful antibiotic that was used to prevent the outbreak of epidemics among slaves during the construction of the pyramids. Herbal therapies and the setting of bones constituted only a small part of the elaborate healing rituals performed by Egyptian priests, who believed illness was caused by malevolent spirits that could be appeased only by magic.

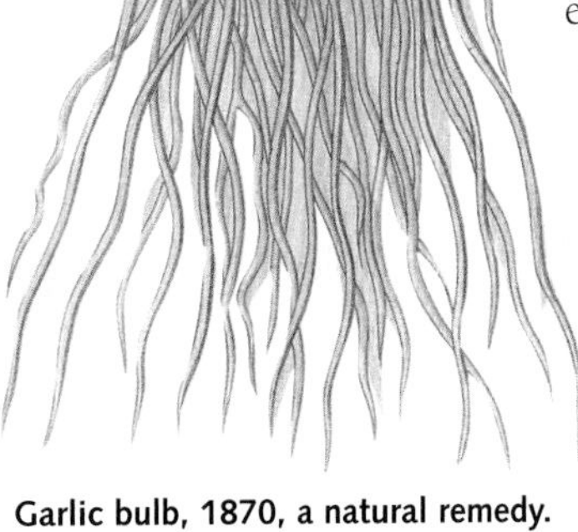

Garlic bulb, 1870, a natural remedy.

Ancient Greeks 600 BC–AD 522

It was the discovery of iron that encouraged the spread of civilization to new areas beyond the traditional valley plains favoured by the great cities of the Bronze Age. Although it is thought that humans first discovered and extracted native iron from meteorites, the key breakthrough came from the Greeks, who mastered the technique of extracting iron from its ore by heating it together with charcoal in a furnace. Unlike that from meteorites, the iron present in ore was in abundant supply, forming a cheap source of metal for the first time in history. It was vastly superior to stone, being more durable and easier to fashion into tools and weapons. Using iron tools, it was possible to clear trees and rocks from vast expanses of land that had previously been unsuitable for agriculture. Tree-felling provided the timber for shipbuilding, which expanded trading opportunities, and new, smaller cities started to spread across Europe.

The Ancient Greeks were beneficiaries of the technological advances of the Iron Age.

They created independent city states, governed by the principle of democracy, and placed the greatest value on study and learning. They extended their own knowledge by absorbing that of countries with which they had contact through trade or war. It is in their search for more knowledge and, especially, in their attempts to explain the world about them that we see the origins of rational science.

Not satisfied with the mysticism favoured by other ancient cultures, some Greeks believed that philosophy — logical arguments grounded on facts that everybody could verify by observation or mathematics — would give them a far superior understanding of the world. This change of approach encouraged a more abstract interest in nature that had less to do with solving immediate practical problems and was more concerned with constructing arguments to explain phenomena such as the behaviour of planets. Most importantly, it allowed the Greeks to pose questions that opened the door to entirely new fields of investigation and discovery, including the study of natural history. What is the relationship between animals, plants and humankind? Why do volcanoes erupt? What causes people to become ill?

The Greek scholars called themselves 'natural philosophers' a practice that would continue until the English scientist, mathematician, philosopher and poet William Whewell (1794–1866) coined the more familiar term 'scientist' in 1840. Two Ancient Greek natural philosophers, Aristotle (384–322 BC) and Hippocrates (*c*. 460–370 BC), deserve special attention because they were particularly influential in shaping our understanding of the natural world.

Arrowheads from the Iron Age.

ARISTOTLE

Greek philosophers were never entirely successful in separating fact from fiction. Aristotle, for example, believed the world was eternal and that plants possessed a 'psyche', a crude version of the human soul. He did, however, establish the tradition of studying animals by dissection, and was the first person to observe and record the development of embryos. His *De Anima* was a triumph of order over a potentially chaotic jumble of information, providing descriptions and observations of over 500 species, including 120 different kinds of fish and 60 of insects. Aristotle believed that nature could be organized along a scale ranging from the most simple, minerals, to the most perfect,

Aristotle. Such images reflected the rediscovery of classical philosophers that helped create the Renaissance.

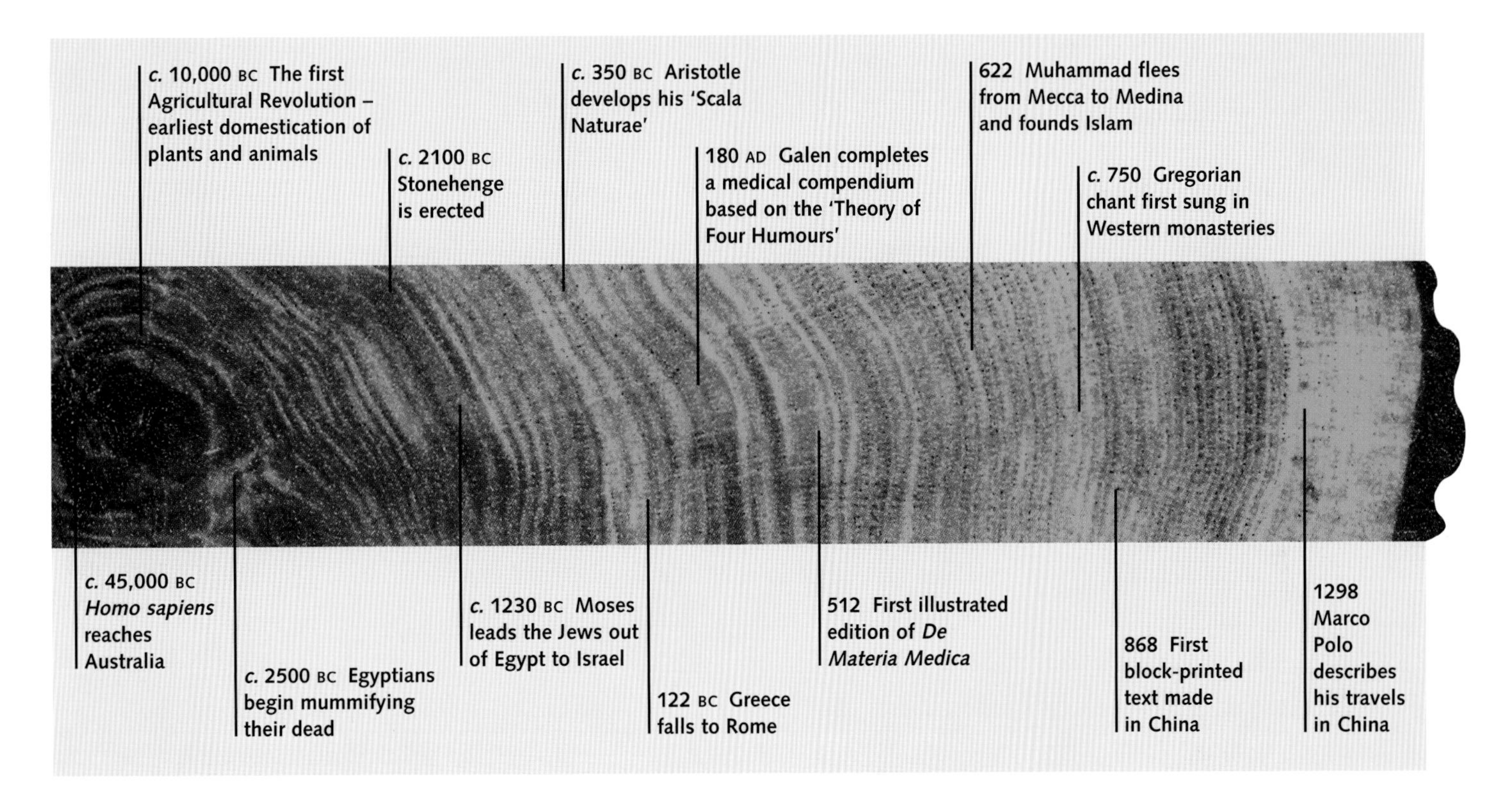

man. This 'scala naturae' was based on the assumption that species were fixed and unchangeable. Aristotle's scheme, which lent support to the Christian belief that God had created every species of plant and animal exactly as we see them today, continued to dominate natural history until the concept of evolution began to gain popularity in the 18th century.

Like all great philosophers, Aristotle had a wide range of interests: although he was primarily interested in plants and animals, his curiosity extended to astronomy and the structure of the Earth, too. In his book *Meteorologica*, he suggested that volcanoes and Earthquakes were the result of winds that blew beneath the surface of the Earth, and that rivers were produced by rain, condensed by mountains from water in the atmosphere. Aristotle argued that fossils were proof that parts of the world had once been covered by water, and that the world was divided into zones dictated by the amount of heat they received from the Sun. The equatorial zone was too hot and the polar zone too cold to support life.

By 500 BC, we can already see the seeds of some of the ideas that would lead to separate fields of study in the 19th and 20th centuries: systematics, comparative biology, ecology and evolutionary biology.

HIPPOCRATES

Medicine is a specialized area of study, separate from natural science, but it has always maintained links with botany, because of the medicinal properties of plants. As recently as the 18th century, all the leading botanists, except John Ray, were physicians, and much of the initial research in biology in the 19th century sprang from medical schools, such as those at Edinburgh and Glasgow.

Californian redwood trees or giant sequoias are the largest and longest living plants. This slice is taken from a tree estimated to have lived for over 1300 years. The timeline shows natural history events in the context of other milestones from 45,000 BC to 1298.

It was the Greek scholar Hippocrates who was responsible for changing medicine from a mystical ritual to a rational practise, based on sound scientific principles. He argued that far from being an act of vengeful Gods, illness was caused by natural factors that could be understood and treated by logical means. Health depended on the balance of the four fluids, or 'humours', that made up the human body: blood, yellow bile, black bile and phlegm. The 'Theory of the Four Humours' stated that if the four fluids remained in equal proportions within the body the person remained well, but any imbalance would cause the person to become physically or mentally sick. The body was believed to have its own vitality, a self-healing process that sought to correct any imbalances. The task of the physician was to use herbs, purging and bloodletting to encourage this natural process.

Hippocrates.

THE FIRST HERBALS

A detailed knowledge of the therapeutic properties of herbs was an important weapon in the physician's arsenal against disease, and by the first century AD both Chinese and European herbalists were compiling their knowledge in the first in a long series of herbal manuscripts. In AD 77, Dioscorides (*c.* 40 – *c.* 90 AD), a Greek physician to Nero's Roman army, compiled his book *De Materia Medica*, including information on over 500 herbs. It was copied by hand until the invention of printing in the 15th century, and became the standard medical reference in Western medicine. Dioscorides gathered his information during extensive travels with the Greek army in Asia. He describes the use of henbane (*Hyoscyamus niger*) to induce sleep and relieve pain, juniper (*Juniperus communis*) to ward off the plague and greater burdock (*Arctium lappa*) to relieve gout, fevers and kidney stones.

The Decline of Natural Philosophy

Although natural philosophy was a dazzling intellectual achievement, which would play an important role in the development of modern science in the 17th century, it had serious limitations. Its greatest flaw was the assumption that human logic alone would explain the natural world. Hippocrates' theory of the four humours and Aristotle's scale of nature, for instance, seemed to provide perfectly logical explanations of the causes of illness and the relationship between plants, animals and humans, and, according to Greek philosophy, this meant they were true.

The dangerous belief that the Greeks had solved many of the puzzles of the Universe stifled further investigation and effectively halted the progress of scientific knowledge in Europe until the Renaissance. For this reason, works such as Dioscorides' *De Materia Medica* and the compilation of all known medical knowledge based on the theories of the four humours, collected by the Greek physician Galen (*c.* 130 – *c.* 201 AD) and published in

Page taken from an illustrated edition of Dioscorides' *De Materia Medica*.

AD 180, were copied over and over, and used without question for the next one and a half thousand years.

Defeat by the Romans led to a further decline in the study of natural philosophy. Although Greek science was widely acclaimed, it produced little in the way of immediate practical application. Improved astronomical observations and geometry led to more accurate maps and the development of a stunning new style of architecture, but at the grass-roots level, in areas such as agriculture and manufacturing, the way of life changed little from that of the Bronze Age 2000 years earlier. The practical Romans were unimpressed, and although they did nothing to actively discourage the study of natural philosophy, they did little to encourage it either.

The economy of the Roman Empire, like that of the Greek Empire before it, was built on the cruel but highly successful principle of invasion, occupation and exploitation of foreign lands. At their prime, the Romans controlled the largest state the world had ever seen, but ultimately there was a limit to the number of lands they could conquer and the number of people they could force into slavery, their principal source of wealth. Spiralling inflation undermined the economy and finally led to the collapse of the Roman Empire.

The Middle Ages (AD 476–1440)

THE MAGIC OF NATURE AND THE GLORIFICATION OF GOD

The collapse of the Roman Empire led to great changes in both political and religious power. One of the most significant was the rise of medieval Christendom, whose influence on Western science was to last until the 13th century. During this period, many European centres of learning were confined to abbeys and monasteries, so priests and clerics became the guardians of intellectual pursuits, maintaining their existing store of knowledge, though rarely adding significantly to it. However, there was a negative side to this guardianship. The Church associated the works of the Ancient Greeks with the sinful pursuit of knowledge for personal pleasure and with the taint of Roman idolatry, and discouraged their study by those outside religious orders, resulting in stagnation of scientific knowledge.

Despite the general clampdown on Greek philosophy, selected scholars continued to study the works of Galen and Dioscorides. The care of the sick remained a central concern of the Church community, which is why these Ancient Greek books continued to dominate medicine throughout the Middle Ages. Christian scholars combined ancient wisdom with that of folklore, a blend of practical knowledge of the properties of locally grown herbs, passed down through the generations. This folklore revolved around the supposed magical properties of plants, animals and minerals. Mushroom rings, for example, were thought to be the sign of fairies dancing in a circle by moonlight, and the roots of the mandrake were believed to contain evil spirits that would cause the plant to scream if it was pulled from the ground.

Despite this clampdown on natural philosophy in northern Europe, scholarship continued to flourish in the Mediterranean region and the Middle East. The Arabs not only translated the work of the Ancient Greeks, but also drew on expert herbal and medicinal knowledge from India and China, adding this to their own discoveries. The Persian philosopher and physician Avicenna (980–1037), for example, discovered the means of distilling essential oils from flowers for use in treating a variety of ailments. Islamic medicine spread throughout the Moslem Empire, which reached from North Africa to Spain, Italy and Portugal. The Moslems built a number of outstanding hospitals and medical schools. The school at Salerno, for example, encouraged men and women of all faiths to study herbal medicine and train to become physicians.

RECONCILIATION

Although it took a while for European feudalism to become established, living conditions improved steadily through the Middle Ages, and the pessimistic fears for the end of the world were gradually replaced by a more hopeful attitude. Surplus wealth was used to fund the development of new cities, and the first Christian universities, including Paris in 1160 and Oxford in 1167.

As the understandable superstitious paranoia of the early Middle Ages faded, Christian scholars became more open to the ideas of the ancient natural philosophers. This changing attitude encouraged the beginnings of a revival in Europe, and scholars started to study the Arabic translations of philosophers such as Aristotle, searching for ways to reconcile his rational explanations of nature with religious doctrine.

These medieval scholars argued that Aristotle's scale of nature supported the Christian concept of the great universal chain of being. The chain linked imperfect man, at the bottom of the scale, to Earth at the centre of the Universe, and eventually to God, who represented the highest pinnacle of perfection. The Dominican cleric Albertus Magnus ('Albert the Great', also known as Doctor Universalis, *c.* 1200–1280) was perhaps the most active of the new breed of medieval naturalists. He wrote extensively on plants, animals and minerals, and ridiculed the fantastic bestiaries, books written during the early Middle Ages that mixed descriptions of real animals and mythical monsters with moral lessons. Albertus wrote one of the first comprehensive works on the study of minerals and, unlike many others at the time, did not believe that fossils were left by Noah's Flood, but preferred to view them as nature's unsuccessful attempts to spontaneously generate living animals from rocks.

These efforts to reconcile religious dogma with ancient science were more about finding evidence to support the Christian interpretation of the world than an interest in understanding nature itself. A true revival of European interest in science and natural history would have to wait until 1453, when the Turkish invasion of Constantinople (present-day Istanbul)

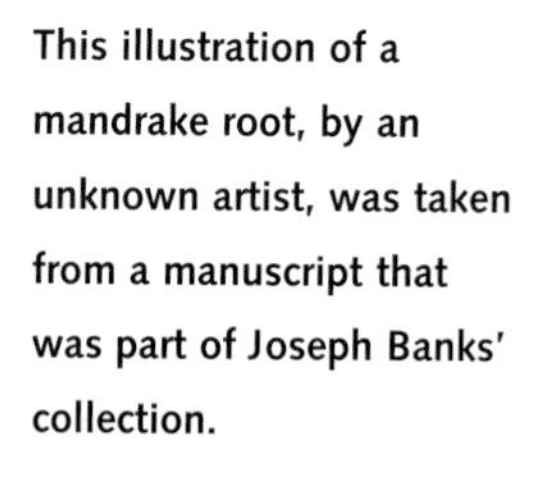

This illustration of a mandrake root, by an unknown artist, was taken from a manuscript that was part of Joseph Banks' collection.

encouraged scholars to flee to the West. They escaped carrying the original works of the ancient Greeks, thereby returning the knowledge censored by the scholars of the early Middle Ages and effectively lost to Europe for over a thousand years.

The Scientific Revolution 1453–1690

Although the original works of the ancients encouraged a revival of interest in natural philosophy, this was only the beginning of a chain of events that led to the birth of modern science two hundred years later. The scientific revolution was a complex affair that involved the development of a new experimental scientific method and its triumphant application by the great English physicist and mathematician Isaac Newton (1642–1727). Looking at the major political, economic and religious changes that swept through Europe, from the beginning of the Renaissance to the end of the 17th century, will help to explain why it occurred when it did — and why the study of nature had to wait until the 19th century to be considered a science like physics and astronomy.

Historiated (decorated) initial 'D' depicting an apothecary, from *The Natural History of Pliny the Elder.*

THE RENAISSANCE — THE SCIENTIFIC REVOLUTION BEGINS (1453–1540)

Technical improvements in the late Middle Ages, mostly in agriculture and cloth production, created a surplus of goods. This triggered a period of rapid economic expansion, and marked the beginning of the Renaissance in Europe. Although a few cities had already introduced the system of exchanging goods for money-payments in the 12th century, the change became more widespread when advances in navigation and shipping improved the availability of the growing surplus of goods.

By the 1400s, most of these goods were being shipped from Asia to Venice, through Germany and on to Britain. With a virtual monopoly on land-based and ocean trading routes, the major Italian cities Venice, Genoa and Milan were the first in Europe to achieve economic and political independence from the Catholic Church. The Church, continuing to benefit from their generous donations, was willing to turn a blind eye to the shifting balance of power.

The new merchant princes established courts that provided an important source of patronage for scholars and intellectuals. Increasing wealth and freedom from the controlling influence of the Church encouraged the growth of the flourishing intellectual and artistic communities from which the Renaissance movement flowered. A revolutionary movement that sought to make a clean break with the medieval past, it favoured

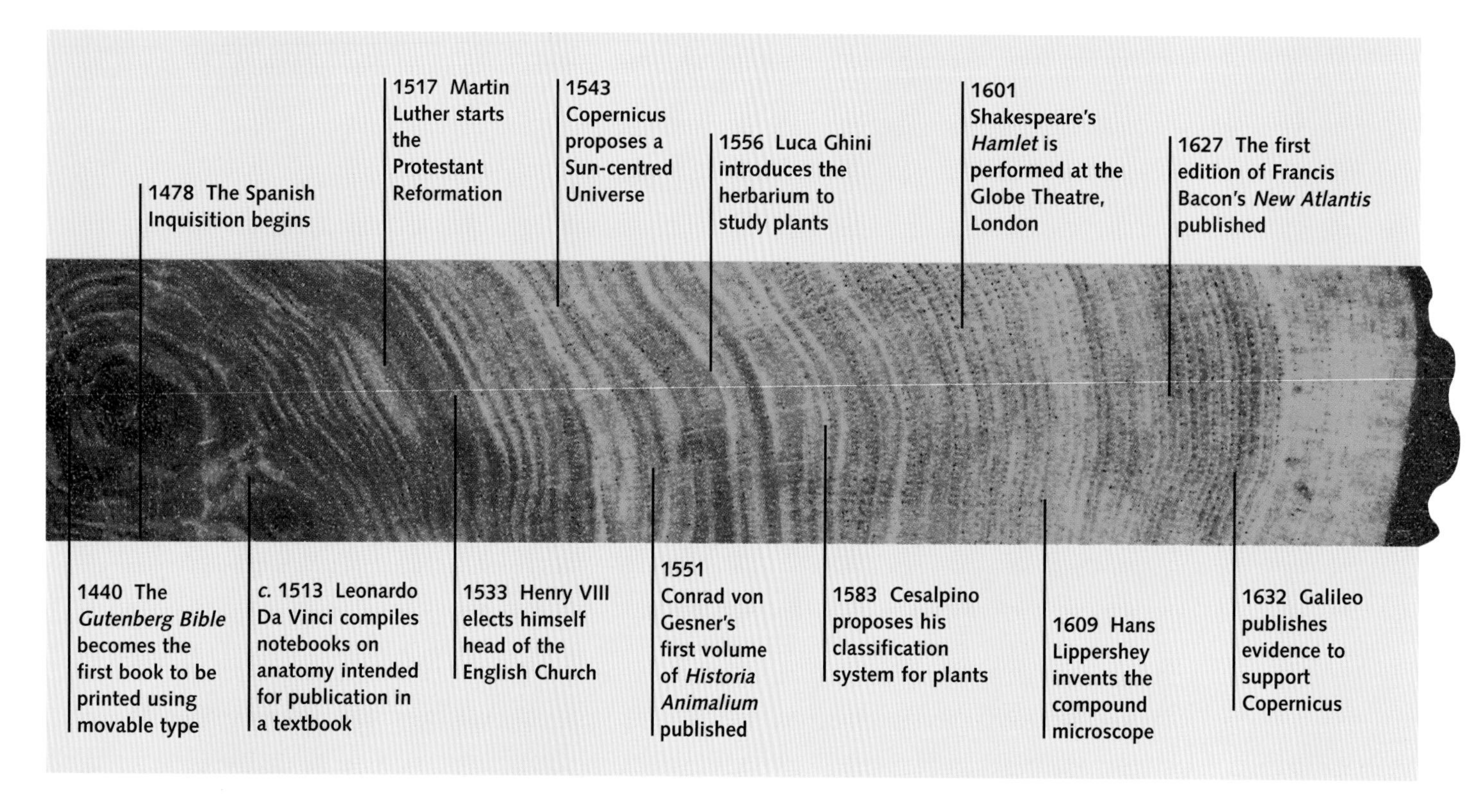

Natural history events and other milestones, 1440–1649.

capitalism over feudalism and created a new culture that encouraged a revival of interest in natural philosophy, as well as classical art and literature.

The movement gradually spread to Germany and the rest of Europe, where it adopted more religious overtones. In Germany, it led to the Protestant Reformation, spearheaded by Martin Luther in 1517, and a nationwide rejection of the Catholic Church.

At the heart of the religious reformation lay the belief in personal judgement and the right of every individual to take responsibility for their own destiny. Such revolutionary ideas eventually led to violent social disorder, with peasant revolts in Germany, Hungary and Spain. In Britain around 1640, and in France around 1790, the movement went a stage further, with civil wars that led to the creation of political democracies and made governments accountable to the people that had elected them.

THE CLASSICS

The rediscovery of the Ancient Greek classics at the beginning of the Renaissance played an important role in the move to create a new culture, free from the influences of the Middle Ages. The humanists set about reading and translating the Greek texts themselves, abandoning the second-hand versions taken from Arabic translations or Latin versions prepared and censored by medieval scholars. The revival of classical learning was to provide an important new framework, from which the 15th and 16th century naturalists could begin to explore the world for themselves.

True to the progressive spirit of the times, the humanist scholars sought to extend the legacy of classical knowledge by adding their own observations to those made by the ancient philosophers. For example, the Swiss naturalist and physician Konrad von Gesner (1516–65), in his *Historia Animalium*, written between 1551 and 1558, attempted to summarize all zoological knowledge gained since the work of classical scholars. His compendium included contributions from fellow naturalists, his own first-hand observations and descriptions of a number of mythical beasts that had infiltrated natural history during the Middle Ages. He listed the animals in alphabetical order, and made no attempt to classify them.

Botany remained the province of medicine and naturalists set about producing updated editions of Dioscorides' works, identifying all the plants for themselves while adding their own more accurate descriptions. The arrangements remained alphabetical, and the descriptions of plants included medical and other practical applications, often with links to the doctrine of signatures, which was thought to reveal their magical properties.

John Dee.

THE DOCTRINE OF SIGNATURES

In isolation, this combination of magic and science might seem like superstitious indulgence, but in reality it presented a common view of the world, shared by the majority of humanist scholars during the 15th and 16th centuries. Plants, animals and minerals had a spiritual and symbolic significance, which was considered every bit as important as their physical appearance in helping to explain the mysteries of nature. This meant that the scholars of the time could hold apparently contradictory beliefs. John Dee (1527–1608), for example, was not only a leading Elizabethan mathematician and astronomer, but also a magician and an astrologer. Magic and science were indistinguishable to the humanists. Both presented a way of helping them to gain a better understanding of the world, and until hard evidence gathered towards the end of the 17th

DOCTRINE OF SIGNATURES

'By the outward shapes and qualities of things we may know their inward vertues, which God hath put in them for the good of Man'.

From a 17th century herbal, thought to have been a translation from the German alchemist and physician Paracelsus (1493–1541)

No-one knows how humans first came to understand the therapeutic properties of plants, but gruesome trial and error must have played an important role. During the Renaissance, many herbalists made a living from selling their herbs to the sick. They blended knowledge gained from centuries of observation with ancient superstitious beliefs in the magical properties of plants. 'The Doctrine of Signatures' as it became formally known, stated that plants had been put on Earth to help cure human illness, and that they carried secret signs or symbols that revealed their therapeutic purpose. For example, walnuts, because of their resemblance to brains, were thought to cure brain disease, whereas henbane, with its rows of fruit that look like teeth, were believed to cure toothache.

CASH CROPS

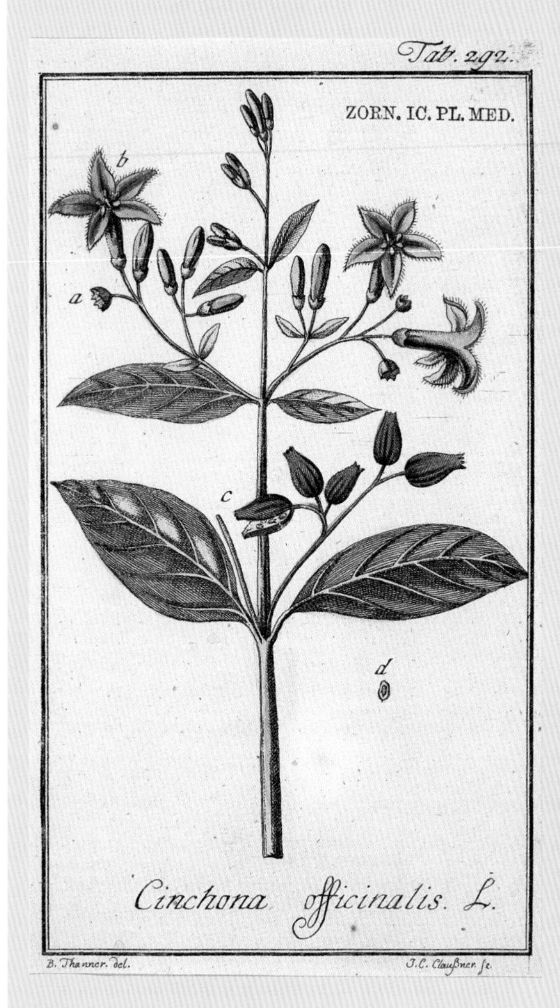

Cinchona officinalis **tree, source of quinine.**

White pepper is one of a succession of plants, including sugar cane, tea, potato, cotton and coffee, that have profoundly influenced the world economy. In 1638, the discovery that quinine, an extract from the bark of cinchona trees, alleviated the problem of malaria helped Europe, Asia and Africa to control a disease that was every bit as devastating as the plague. Quinine enabled white people to build vast empires. It made the mass transportation of slaves and settlement in previously uninhabitable areas such as India and the West Indies financially attractive, and world demand soon outstripped natural supplies. Britain and Holland partly resolved the problem by cultivating Peruvian cinchona trees on plantations in India, but German chemists found a more profitable answer in the shape of synthetic substitutes. Not only did quinine provide the means by which nations could build empires, but it also triggered the development of coal-tar chemistry, a completely new industry, and the production of compounds such as aspirin, saccharin, nylon and PVC. The first quinine substitutes started to appear in the 1930s, and were consumed by millions of men and women during World War II. Without quinine substitutes to protect the allied forces in the Pacific and the Mediterranean, the outcome of the war might have been very different indeed.

century began to prove otherwise, both offered the potential to help them profit from nature.

VOYAGES

Overseas trade had already become an important source of income, and 15th century merchants were eager to expand business. They invested heavily in ship-building, and commissioned astronomers, mathematicians and geographers to help them find new trading routes.

By 1470, Turkish interference with the overland trade routes that ran between Asia and Italy made mastery of the uncharted seas even more critical. Venice had long enjoyed a virtual

Vasco da Gama, by Pedro Barretti de Resende, 1646.

monopoly on the trade of spices, with white pepper accounting for more than half of its imports. Pepper was an extremely valuable commodity, ten times more expensive than any other spice. In the 15th century, salt was the most widely used method of preserving meat, and pepper made the end result palatable. Pepper was a particularly important commodity for sailors who relied on a diet of salted meat at sea. It could be argued that the Portuguese, Italian and Spanish quest to establish new ocean routes to Asia was motivated as much by the need for pepper as by curiosity or economics.

It was the successful combination of classical astronomy, Arabic trigonometry and the practical wisdom of seasoned Atlantic sailors that made the great voyages of discovery possible. These diverse inputs resulted in critical improvements to maps, and accurate astronomical tables that were simple enough for sailors to use at sea. Navigating from these new tables and Gerson's cross staff (a surveying instrument consisting of a staff topped by a frame containing two pairs of sights at right angles), the Portuguese mounted successful explorations around the coast of Africa. In 1498, the Portuguese navigator Vasco da Gama (*c.* 1469–1525) became the first explorer to reach India via the Cape of Good Hope.

Seven years earlier, the Italian navigator and explorer Christopher Columbus (1451–1506), working in the service of the Spanish king and queen, had travelled west, straight out across the Atlantic Ocean, in search of China on the other side of the world. Ignoring the popular beliefs that his voyage would last forever, or that the ship might plunge over the edge of the world, Columbus found the West Indies in 1492. He returned to a hero's welcome, bearing gold, cotton, bananas, various dried and living plants, live parrots, and a rich supply of dead bird and animal specimens that had never been seen in Europe before.

Naturalists such as Oviedo y Valdes (1478–1557) set about describing these new species in his *History of the Indies*, but the volume of specimens and the inaccuracies of the ancient Greek texts soon drew their attention to the inadequacies of the classical system of description and organization. The colonization of America and the Indies attracted more practical men too — farmers and miners, who sought to make a living from the abundant supply of natural resources. These men both wrote and referred to a growing number of practical works, including *De Re Metallica*, a guide to the identification and classification of minerals and the techniques and economics of metal and mineral mining. Published in 1556, this important work was written by the German mineralogist and metallurgist Georg Bauer, better known as Georgius Agricola (1494–1555), and was based on his practical experience in Saxon mineral mining. Guides such as *De Re Metallica* contributed as much to the understanding of nature as did the natural philosophers of the day.

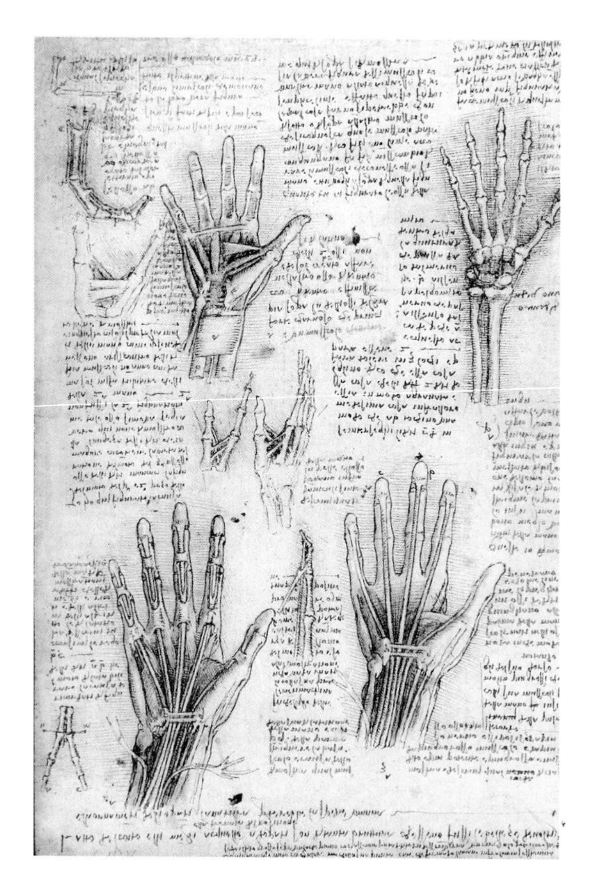

Anatomical sketch by Leonardo da Vinci.

ART

For the first time in history, art during the Renaissance was valued as a commodity. Artists were much in demand by the new merchant princes, who commissioned works as investments and adornments to illustrate their affluent lifestyles. Handsomely compensated for their efforts, Italian artists set up city studios alongside the new universities and laboratories. The subsequent cross-fertilization of thoughts and ideas was to have a stimulating effect on the development of science.

Two contributions from Renaissance art — the invention of perspective and the triumph of 'realism' over classical idealized form — changed the way people viewed the world, and played a vital role in influencing the progress of natural history. An understanding of perspective enabled artists to represent three-dimensional figures in two dimensions, a conceptual leap that paved the way for new realistic paintings. These are best characterized by the work of artists such as the Italian Leonardo da Vinci and the German Albrecht Dürer.

Da Vinci was the original Renaissance polymath: artist, musician, engineer, scientist and mathematician, he was a man of many talents. Not satisfied with painting an accurate representation of what he saw, he wanted to understand the underlying nature of the subjects he painted and, if they moved, the 'mechanism' by which the movement was produced. To paint a human, for example, required detailed observations of anatomy and a study of the muscles, organs and skeleton beneath the skin. Not content to rely on the teachings of Galen and Hippocrates, Leonardo conducted his own dissections, and in about 1523 created a series of stunningly accurate anatomical drawings in notebooks that he planned to publish as a textbook. The drawings were never used for the purpose he intended, but they did provide undeniable truth that the anatomical teachings of Galen and Hippocrates were incorrect.

Study of the wing of a hooded crow, painted on vellum with gold, by Albrecht Dürer, 1512.

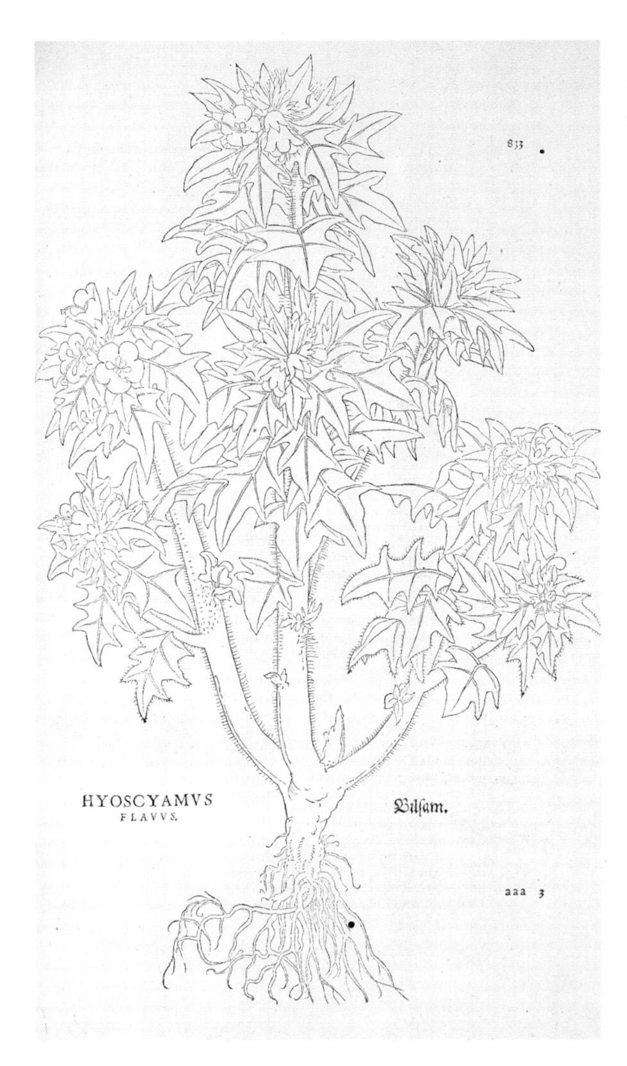

LEFT: **Illustration of henbane from the 15th century.**

floures of Mandrake there is no ſuch delectable or amiable ſmell as was in theſe amiable floures which *Ruben* brought home. Beſides, we reade not that *Rahel* conceiued hereupon, for *Leah Iacobs* wife had foure children before God granted that bleſſing of fruitfulneſſe vnto *Rahel*. And laſt of all, (which is my chiefeſt reaſon) *Iacob* was angry with *Rahel* when ſhee ſaid, Giue mee children or els I die; and demanded of her, whether he were in the ſtead of God or no, who had withheld from her the fruit of her body. And we know the Prophet *Dauid* ſaith, Children & the fruit of the womb are the inheritance that commeth of the Lord, *Pſal*. 127.

Serapio, *Avicen*, and *Paulus Ægineta* write, That the ſeed and fruit of *Mandragoras* taken in drinke, do clenſe the matrix or mother: and *Dioſcorides* wrot the ſame long before them.

He that would know more hereof, may reade that chapter of Dʳ *Turners* booke concerning this matter, where he hath written largely and learnedly of this Simple.

CHAP. 66. *Of Henbane.*

1 *Hyoſcyamus niger.* Blacke Henbane.

2 *Hyoſcyamus albus.* White Henbane.

¶ *The Deſcription.*

1 THe common blacke Henbane hath great and ſoft ſtalkes, leaues very broad, ſoft, and woolly, ſomewhat jagged, eſpecially thoſe that grow neere to the ground, and thoſe that grow vpon the ſtalke, narrower, ſmaller, and ſharper, the floures are bell-faſhion, of a feint yellowiſh white, and browne within towards the bottome: when the floures are gone, there

Gg

RIGHT: **Illustration of henbane from the late 16th century.**

Naturalists gradually followed the lead set by da Vinci and Dürer, and by the middle of the 16th century numerous books were published, containing realistic illustrations of plants, animals and minerals. The quality of these illustrations gave naturalists more accurate information than any wordy description. The process of development can be seen in the example above, which compares woodcuts, illustrating the poisonous medicinal plant henbane, taken from herbals in the 15th century and the latter half of the 16th century.

In 1543, the Belgian anatomist Andreas Vesalius (1514–64) published *De Humani Corporis Fabrica*, and took the process a stage further — none of the information in this work was reproduced from ancient texts, being based instead on first-hand observation of dissections.

CLEARING THE DECKS

The triumph of navigation, more than any other development during this first phase of the scientific revolution, secured the fate of science and proved that it could be applied for profitable advantage. Religious controversy had challenged orthodox beliefs and given people the confidence to think for themselves and rely on their own judgements. In natural history, this encouraged scholars to make their own detailed observations of plants, animals and minerals. Improved accuracy of illustration and the steady flood of new species arriving from the colonies, highlighted the limitations of systems used by the ancient Greek naturalists. The humanist scholars weren't able to suggest an answer to these problems, but they did set the agenda for the developments that would follow in the next century.

The Rise of the Scientific Method 1540–1650

When Britain and Holland took over from Spain and Portugal as the leading trading nations in Europe, the flourishing new merchant classes began to campaign for more political power and an end to the old system of feudal control. It was a long, hard struggle. The resistance of the kings to such progress resulted in violent revolts and open warfare, but in 1576 the democratic States General of Holland and in 1649 Cromwell's Commonwealth of England were born.

Despite protracted civil unrest, Holland and England continued to flourish. Unlike their feudal rivals Spain and Portugal, their strength lay in their natural resources and the ability to combine manufacturing, mostly woollen cloth-making, with trade.

The collapse of the feudal system in England and Holland created a pool of cheap labour, reducing the costs of goods at a time when booming overseas trade was creating increased demand and prices. The profits were unprecedented, shipbuilding and the production of cast iron increased, and the demand for raw materials (particularly timber for iron smelting but also pitch, flax and corn) stimulated the economic growth of the Baltic countries of Sweden, Denmark, Poland and Russia.

Portrait of Galileo Galilei, by Ottavio Mario Leoni, (*c.* 1578–1630).

A NEW EXPERIMENTAL APPROACH

It was against this background, of democracy, the rise of capitalism in Northern Europe, and the search to find new ways of solving the challenges created by the growth of industrialization, that the framework of modern science fell into place. Mining and the discovery of cash crops, such as sugar and tobacco, during the Renaissance had shown that there were profits to be made from nature, and highlighted the inadequacies and limitations of the classical interpretation of the world. The challenge was to create new ways of analysing nature that were more in keeping with the progressive spirit of the age, and to move beyond the teachings of the classical scholars in a search for new ways of extending knowledge. It was the successful use of new experimental methods in astronomy and physiology that would provide the key.

REVOLVING AROUND THE SUN

In the year of his death, the Polish astronomer Nicolaus Copernicus (1473–1543) published his only scientific work, *De Revolutionibus Orbium Coelestium*, presenting the revolutionary suggestion that the Earth was a revolving sphere and the Sun sat at the centre of the Universe. His theory challenged the view held by the Catholic Church since the Middle Ages of a closed Universe, with the Earth at the centre surrounded by concentric spheres that included the Moon, Sun, planets and stars, in a succession of perfection. When Copernicus published his theory, it was impossible to say with any certainty that the Sun-centred view of the Universe offered a more accurate model of the world than the Earth-centred version. To have any credibility, Copernicus needed a means of accurately describing the orbits of the planets and a convincing argument to explain how the Earth could be revolving continuously without hurling objects into space.

It was the brilliance of the German astronomer Johannes Kepler (1571–1630), who suggested that the planets followed elliptical rather than circular orbits, and the perfection of the refracting telescope by the Italian astronomer, mathematician and physicist Galileo Galilei (1564–1642) that supplied the evidence. Not content to rely on his more accurate observations of the galaxy, Galileo went an important step further and formulated a mathematical description of the motion of bodies to provide an argument to explain the motion of the Earth. He did this by conducting exhaustive experiments, and making accurate measurements of the rate of falling bodies, using slopes and pendulums. In 1632, Galileo published his evidence in his *Dialogue Concerning the Two Chief World Systems, Ptolemaic and Copernican*.

The Catholic Church banned the book, and Galileo was forced to recant his views before the Inquisition in 1633. Despite the negative outcome of the trial, most scientists, especially those in Protestant countries that had abandoned their allegiance to Rome, deemed Galileo's defence a triumph for the new experimental science over religious dogma.

A replica of Galileo's refracting telescope.

THE CIRCULATION OF BLOOD

'I profess to learn and teach anatomy, not from books but from dissections, not from the tenets of Philosophers but from the fabric of nature.'

William Harvey (1578–1657)

The excitement created by Galileo's experimental science inspired the English physician William Harvey to take an experimental approach to human anatomy and physiology, in an attempt to try and explain the mechanics of the movement of blood around the body.

In marked contrast to the traditional, observation-based approach to anatomy, Harvey's *Exercitatio Anatomica de Motu Cordis et Sanguinis*, published in 1628, reads more like a piece of modern scientific research. It includes a description of a series of practical experiments to demonstrate the mechanics of blood flow, based on an analogy between the body and a hydraulic machine. Harvey couldn't prove the existence of capillaries, the fine hair-like tubules that link the supply of blood between veins and arteries. They were too small to see with the naked eye, but he did predict their existence. It was the Italian naturalist Marcello Malphigi (1628–94) who was the first naturalist to observe capillaries, in 1660, and provide overwhelming evidence to support Harvey's theory. Later work revealed that the circulation of the blood was more complicated than the system Harvey had described, but his carefully constructed evidence was readily accepted by his contemporaries, because it offered a far more accurate explanation than the classical theory that had dominated medicine for thousands of years.

Harvey's experimental approach was revolutionary, because it demolished the notion of a body controlled by vital spirits in favour of one governed by entirely natural processes that could be understood and, most importantly, controlled. It raised new questions and opened up a whole new field of research for scientists to explore. Where would our understanding of physiology and modern medicine be today without an understanding of the blood circulatory system?

Francis Bacon, by William Marshall, 1640.

A PIONEER OF CLASSIFICATION

Naturalists in the 16th century may not have developed experiments to advance their understanding of plants, but they did come up with new methods of studying both living and preserved specimens. In 1542, the first botanical gardens were created in Padua, and in 1556 the Italian botanist Luca Ghini pioneered the use of the herbarium (a collection of dried and pressed plants) to study the structure of plants. Botanists continued to struggle with the volume of information created by the ever-growing variety of plants arriving from the New World. However, there was little attempt to describe them in anything but the traditional way: by alphabetical order, practical benefit to humankind and magical properties.

The Italian physician and botanist Andrea Cesalpino (1519–1603) proved the exception to the rule when he took the pioneering step of creating a classification system for plants based exclusively on their biological characteristics. He borrowed the idea from Aristotle but, in the true spirit of humanism, he went a stage further than the ancient philosopher, and tried to establish the essential characteristics that would help identify the 'essence' of each species. He decided that nutrition and reproduction were the most important functions, and set about arranging his plants into groups by comparing the structure of their roots and flowers. In 1583, Cesalpino published his findings in his book *De Plantis*, but because his system was based on the already discredited ideas of Aristotle, he failed to influence his fellow botanists. Cesalpino may not have gained acceptance for his ideas when he was alive, but he was recognized one hundred years later by naturalists involved in the creation of the modern system of taxonomy.

SPIN DOCTORS

'The true and lawful end of the sciences is that human life be enriched by new discoveries and powers.'

Francis Bacon, *New Atlantis*, 1627

It is unlikely that the new experimental approach would have achieved its wide acceptance as the new method of science without the pioneering efforts of two great men: the English philosopher, lawyer and statesman Francis Bacon (1561–1626) and the French philosopher and mathematician René Descartes (1596–1650). Publicists and visionaries, both men argued that all science should be based on the new scientific method, because it offered a far more accurate way of analysing the world than the more speculative philosophy of the Greeks or the magical symbolism of the Renaissance. Although they differed in the precise detail of the method they described, they promoted the same utilitarian belief that it would enable men to conquer and exploit nature for profit and in Descartes' own words 'enjoy without trouble all the fruits of the Earth'.

René Descartes, by Frans Hals.

The new scientific method embodied many of the ideas of the times. It went hand in hand with the religious reformation and the rise of commercialism. The Catholic religion worked on the principle of subordination, the ownership of land and a hierarchy of power whereas the Protestant 'work ethic' encouraged individuals to make their own decisions and their own fortunes. The Protestants believed that wealth was a symbol of spiritual worth. If God had created the world with all of its riches for the benefit of humankind then the practical search for knowledge made the new science a religious activity in its own right.

THE FRUITS OF KNOWLEDGE 1650–1690

The last 40 years of the scientific revolution were ones of relative stability after the violent political and religious reformations of the previous century. These were constructive years for the flourishing new civilizations, with governments and the ruling classes eager to invest in activities that would improve trade, manufacturing and agriculture. By 1650, science had become a prime target for development, and the increase in funding motivated an intense period of activity and growth, concentrated mostly in London and Paris.

These dramatic advances in the acceptance of the scientific method were Bacon's and Descartes' true 'fruits' — fruits of the human mind, rather than of nature. As science flourished and the number of scientists grew, there was a natural tendency for them to gather together to exchange ideas and knowledge. The natural progression of these informal clubs was the creation of scientific societies, such as the French Académie des Sciences, inspired partly by Bacon's vision of scientists 'winning the secrets of nature' by working together in the spirit of collaboration. The founding charter of the Royal Society in England in 1662 was based on a quote from Bacon: 'Further promoting by the authority of experiment the sciences of natural things and useful arts'.

NATURE BEGINS TO LOSE ITS MAGIC

Invented in 1609, the compound light microscope revealed a new level of complexity in nature and an unexpected source of minute creatures for classification. The microscope was used by a number of 17th century naturalists, including the Italian Marcello Malphigi (1628–94), the Englishman Robert Hooke (1635–1703) and the Dutchman Antonie van Leeuwenhoek (1632–1723), who discovered spermatozoa. Some early microscopists, such as Hooke, were particularly fascinated with the anatomy of insects. Before these detailed observations were possible, naturalists believed that flies were particles of flying flesh, spontaneously generated from rotting meat. Their newly discovered anatomy suggested that far from magical pieces of enlivened flesh, flies were complex organisms in their own right.

In 1668 the Italian physician and poet Francesco Redi (1626–97) produced persuasive evidence to support this observation with an experiment that showed that flies and meat were separate entities. By putting a piece of meat in two separate bottles and covering the mouth of one with muslin, he was able to observe flies entering and leaving the uncovered jar. When maggots appeared only on meat in the uncovered jar, Redi argued that flies must make direct contact with meat to lay eggs, and that the eggs, not the meat, produce the maggots.

The generation that believed in the magical properties of nature was drawing to a close. The new rational science and improvements in techniques of observation were starting to provide a more accurate explanation of the natural world and, more importantly, prove that magic served no practical purpose. Redi's experiment heralded the beginnings of our understanding of species as separate, distinct units, an important concept for classification and later for an understanding of evolution. By the early 18th century, alchemists' recipes for the spontaneous generation of organisms, including a delightful description of how to make mice from old underclothes, were destined to become mere curiosities for the museum shelf.

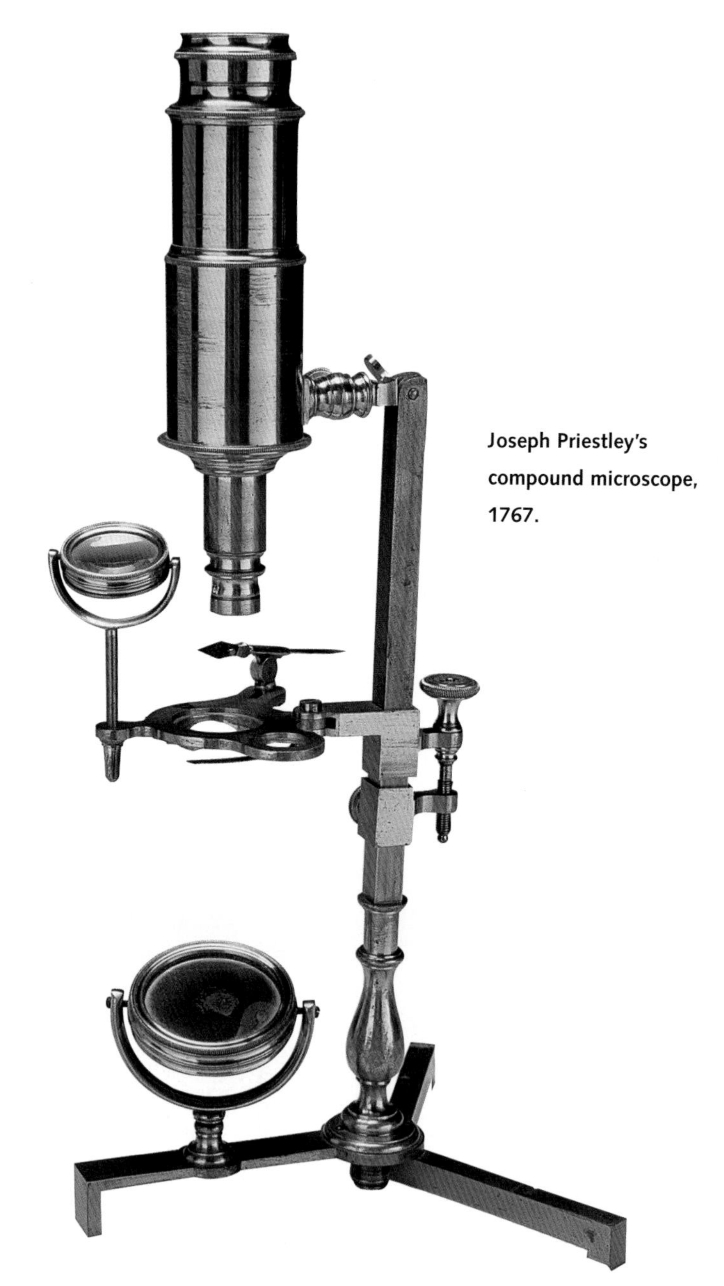

Joseph Priestley's compound microscope, 1767.

FOSSILS AND THE ORIGINS OF GEOLOGY

In 1669, the Danish anatomist, geologist and theologian Nicolaus Steno (1638–86), serving as a physician in Italy, recognized the fossils he collected from rocks in the Tuscan hills as the petrified teeth of a once-living shark. Fossils had been puzzling naturalists for centuries. During the Renaissance, many believed that they were the abandoned efforts of God's unsuccessful attempts to generate life.

Steno found his fossils among broken and irregular layers of rock. What had upset the regular pattern of rock layers, and why were the remains of sea animals found high in the Tuscan hills? Steno believed that his shark and other marine animals had become trapped in sediment at the bottom of the ocean. Successive deposits of sediment eventually formed rock, and the teeth became petrified inside. Steno observed two distinct layers of sedimentary rock in the Tuscan hills. He believed they were the result of two great events in the history of the Earth — the original creation of the planet and Noah's Flood. He suggested that the bumps and folds in the otherwise uniform pattern of the rocks were the result of water damage after each event took place.

Fossilized sharks' teeth: *Jaekelotodus trigonalis* (left), and *Carcharias hopei* (right), from 40-million-year-old rocks, Barton (Hampshire), England.

Although Steno referred to the Bible to explain the origin of the rock formations, fellow naturalists were unwilling to accept that fossils were the remains of extinct animals, or that the Earth's surface was changing shape. Neither explanation fitted with the story of Genesis, in which the Earth, with all its plants, animals, oceans, mountains, valleys and plains, had been created by God, for our human benefit, just as we see them today. Although Steno's ideas were ahead of their time, his observation planted the germ of important ideas that helped to establish geology as a separate field of study in the 18th century.

Noah's Ark, *Nuremburg Bible.*

Newton and Linnaeus Shape Natural History in the 18th Century

NEWTON'S GREAT ACHIEVEMENT

In 1687, Isaac Newton published his greatest work, *Philosophiae Naturalis Principia Mathematica*, which included the proof of his universal theory of gravitation, and marked the birth of science — the grand finale of the scientific revolution. It was one of the most important individual scientific achievements of all time. Newton's laws of gravity and motion described the rules that govern the movement of everything, from apples falling off trees to revolving planets and shooting stars in space. It was an amazing feat, because it took the Copernican system, Kepler's laws of planetary ellipses, and Galileo's dynamics, and explained the common thread between apparently unrelated phenomena in astronomy, physics and mathematics.

Sir Isaac Newton.

Newton's discovery was seen as a triumph of the new scientific method, because it demonstrated that the physical Universe was ordered by simple rules that anybody could understand with careful examination, experimentation and mathematical analysis.

Despite its challenge to the traditional Christian belief that God directly controlled the processes that govern the Earth, Newton's theory became so popular it triggered the beginning of the Enlightenment, a period of political and cultural reform that swept through Europe in the early 18th century and led to the French Revolution. Newton had shown that human reason could find answers to seemingly insoluble problems, and it became fashionable to adopt a 'rational' approach to the social and political issues of the day.

Newton's theory was successful because he was able to argue a convincing case for a combination of rational science and religion. He was a devout Protestant and believed that, far from denying the existence of a God, physics celebrated His power and wisdom. In Newton's vision of the Universe, God was removed to a higher level of control. He created gravity and motion but took no part in their day-to-day running, leaving the door wide open for entirely rational explanations of the forces at large in the Universe. The modern disciplines of physics and astronomy were born when Newton successfully separated his religious beliefs from his rational investigation of the Earth and space.

Although naturalists were inspired by Newton's discovery, another one hundred and fifty years would pass before a further modification in religious belief transformed their investigation of nature into a science. There was no incentive to look for rational explanations of the origins of the Earth, humankind and the diversity of organisms because the Bible stories of the Creation, Genesis, and Noah's Flood continued to offer a more than satisfactory explanation. Naturalists had to accumulate a lot more evidence before they could begin to speculate that the rich variety of life and the amazing geography of the world, with all its oceans, deserts and mountains, was anything other than

Linnaeus in his Laplander costume.

confirmation of divine craftsmanship.

Natural history may not have become a science during the 18th century, but Newton's ordered approach did encourage naturalists to develop new systems of classifying organisms — identifying and grouping species according to their physical similarities and differences. They believed that imposing order on the apparently chaotic world of nature would enable them to find their own evidence of God's master plan in the Universe and, most importantly, make the best practical use of the world He had entrusted to them. The need for classification became increasingly important as a bewildering variety of specimens continued to flood back to Europe from the colonies and voyages of discovery.

LINNAEUS AND CLASSIFICATION

'The most compleat naturalist the world had ever seen.'

Sir William Watson, *Gentleman's Magazine*, 1754

The Swedish-born botanist and physician Carl (or Carolus) Linnaeus, originally Carl von Linné (1707–1778), deserves a special mention in this short history because of his vital role in helping to establish the modern system of biological taxonomy (classification). It is often wrongly assumed that he developed the modern classification system single-handed, but Linnaeus was the first to acknowledge the contributions of others, particularly his close friend Peter Artedi (1705–1735), a brilliant naturalist who died prematurely, aged only 30.

Linnaeus was successful in developing a system that allowed collectors in the field to categorize large numbers of plants and animals more rapidly than had been possible with earlier systems. However, it had serious limitations, and was subject to modification from the start. It was Linnaeus's development of the binomial system for naming species that earned him his lasting fame, and even that — as we will see on p.28 — was more a case of circumstance than conscious design.

Linnaeus's great skills as a botanist, and later his encyclopaedic mission to classify every species on Earth, earned him international fame and a succession of wealthy patrons, from wealthy Dutch businessman George Clifford, the director of the East India Company, to the King and Queen of Sweden. Linnaeus was deeply patriotic, and believed that the discovery of new plants and mineral reserves would reap important economic benefits for Sweden. For example, after becoming a lecturer in botany at Uppsala University, near Stockholm, he conducted the first national survey of Swedish Lapland in 1732. And much later, while professor of medicine and botany at Uppsala, Linnaeus attempted to identify a suitable Swedish plant that could be cultivated as a cheap alternative to the expensive option of importing tea from China.

Linnaeus not only created a convenient method for identifying and classifying large numbers of new specimens, but also established the tradition of sending naturalists on voyages to collect specimens from around the world. His 'apostles' included a number of promising students who risked their lives and, occasionally, as in the case of Pehr Lofling during his exploration of South America, died in the attempt to better understand the flora of faraway lands. These early journeys not only contributed to the accumulated knowledge of natural history in the 18th century, but also paved the way for the heroic voyages that culminated in Darwin and Wallace's revolutionary theory of evolution, nearly a hundred years later.

SEX AND POETRY

Carl Linnaeus and Peter Artedi first began to develop their classification system while studying as undergraduate students at the University of Uppsala. Artedi was keen to find a new way of classifying the Umbelliferaceae, the family of plants that includes cow parsley and hemlock, and it was his interest that first triggered Linnaeus's attempt to do the same for the rest of the plant kingdom.

Artedi and Linnaeus were familiar with earlier systems developed by the Italian botanist Andrea Cesalpino (1519–1603), the English naturalist John Ray (1627–1705), and the French botanist Joseph Pitton de Tournefort (1656–1708), but found them difficult to use. It was a work published by de Tournefort's pupil, Sebastien Vaillant, on the sexual characteristics of plants that gave Linnaeus the idea of developing a new system based on the number and appearance of the stamens and pistils, the male and female reproductive organs of plants.

He took the flowering plants (angiosperms) and divided them into classes according to their male parts, then divided the classes into orders based on the female parts. Linnaeus, a devout Christian with a gift for poetry and a delight in the aesthetics of nature, enlivened his classification system with evocative metaphors.

Natural history events and other milestones, 1660–1799.

1660 Royal Society founded in England

1661 John Evelyn warns of the effects of air pollution

1668 Francesco Redi experiments with flies

1669 Nicolaus Steno recognises fossilized shark's teeth in Tuscany, Italy

1696 Isaac Newton publishes *Principia Mathematica*

1735 Peter Artedi dies

1748 The excavation of Pompeii begins

1753 Linnaeus publishes his *Species Plantarum*

1758 Linnaeus publishes his *Systema Naturae*, 10th edn.

1768 Captain Cook sets out on his voyages to the South Seas

1776 American Declaration of Independence

1789 The French Revolution

1798 Malthus publishes his *Essays on the Principles of Population*

***c.* 1799** The Neptunism versus Plutonism debate begins

He described the class Monandria, for example, consisting of plants with a single stamen (male part), as 'one husband in the same marriage', Diandria, those with two stamens, 'two husbands in the same marriage' and Polyandria, plants with multiple stamens, 'twenty males or more in the same bed with the female'!

As the following quotation shows, Linnaeus had already taken the risky step of equating the sex life of plants with animals in his earlier book *Praeludia Sponsaliarum*:

'The actual petals of a flower contribute nothing to generation, serving only as the bridal bed which the great Creator has so gloriously prepared, adorned with such precious bed curtains, and perfumed with so many sweet scents in order that the bridegroom and bride may therein celebrate their nuptials with the greater solemnity. When the bed has thus been made ready, then is the time for the bridegroom to embrace his beloved wife and surrender himself to her...'

Linnaeus's classification system, however, with its suggestions of incest, polygamy and polyandry, took the analogy one step further, and was considered outrageous by many of his time. Johann Siegesbeck, a scholar from St Petersburg, for example, denounced the system as 'lewd', commenting 'Who would have thought that bluebells, lilies and onions could have been up to such immorality?' Linnaeus didn't respond directly to Siegesbeck's criticism, but he did name an unattractive weed *Sigesbeckia* in his critic's honour!

A FORGOTTEN HERO

Linnaeus summarized his classification system for plants in 1753 in his *Species Plantarum*. He called this work 'the greatest Botany' and listed every plant that he had ever encountered, including the wealth of exotic species that he had catalogued for George Clifford while working in his private botanical gardens at Hartekamp. This first two-volume edition included 5900 species, growing to over 7700 by the time of Linnaeus's death in 1778. Ever since his days at Uppsala, Linnaeus had nursed an ambition to classify every species on Earth, and in 1758 he published a comprehensive classification system for zoology in the tenth edition of his *Systema Naturae*. This system was clearly the result of his earlier collaboration with Peter Artedi.

Towards the end of September 1735, Artedi had spent several hours with Linnaeus, discussing the research that he had recently completed on fish during a visit to England. He planned to publish this work, but tragically, a few days later, returning home from an evening meal, missed his footing and fell into one of the many canals in Amsterdam and drowned. Devastated by his friend's death, Linnaeus raised the necessary funds to publish the book posthumously. Artedi's *The Natural History of Fishes* appeared in 1736, and described a classification system that was identical in every way to the one favoured by Linnaeus. The great botanist made no attempt to disguise Artedi's contribution to the system, and was in fact full of praise for his friend's achievement, remarking 'You will see more perfection than can be expected in botany for a hundred years to come. He has established natural classes (orders), natural genera, complete characters, and a universal index of synonyms, incomparable descriptions, and unexceptionable specific definitions'.

Had he lived, Artedi would probably have shared in the recognition that Linnaeus enjoyed after the publication of his *Species Plantarum* and *Systema Naturae*. We will never know for sure, but having published only a single work, Artedi's contribution was inevitably lost

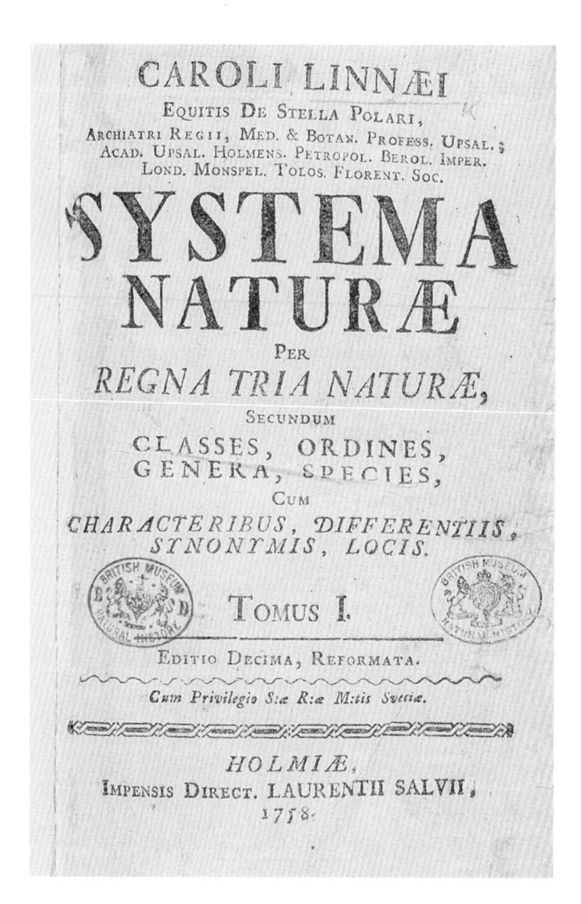
CAROLI LINNÆI
EQUITIS DE STELLA POLARI,
ARCHIATRI REGII, MED. & BOTAN. PROFESS. UPSAL.; ACAD. UPSAL. HOLMENS. PETROPOL. BEROL. IMPER. LOND. MONSPEL. TOLOS. FLORENT. SOC.
SYSTEMA NATURÆ
PER
REGNA TRIA NATURÆ,
SECUNDUM
CLASSES, ORDINES, GENERA, SPECIES,
CUM
CHARACTERIBUS, DIFFERENTIIS, SYNONYMIS, LOCIS.
TOMUS I.
EDITIO DECIMA, REFORMATA.
Cum Privilegio S:æ R:æ M:tis Sveciæ.
HOLMIÆ,
IMPENSIS DIRECT. LAURENTII SALVII,
1758.

LEFT: **Title page of *Systema Naturae*, a compendium and classification of the natural world by Carl Linnaeus in 1758, 10th edition.**

RIGHT: **George Dionysius Ehret's watercolour published in 1736 illustrating Linnaeus's 'sexual system' for the classification of plants.**

amongst the sheer volume of work published by his prolific friend over the next forty years. Although Linnaeus made no attempt to disguise their early collaboration, he didn't actively promote Artedi's name after his death either. Whether this was a conscious decision on Linnaeus's part or an inevitable effect of time, without a champion to keep his name alive, fate ensured that Artedi was destined — like many others before and since — to join the ranks of the forgotten heroes of natural science.

PUTTING A NAME TO IT

Before Linnaeus published his *Species Plantarum* and his *Systema Naturae*, the naming of plants and later animals was a haphazard process that was open to confusion. The lengths of names varied, and became particularly cumbersome when attempting to differentiate individual species within groups of very similar plants. It was impossible to remember the lengthy descriptions of every known plant, and this made identification painfully slow. Linnaeus attempted to address the problem by introducing his system of generic names — short, memorable names to distinguish the major groups or genera of plants.

It was much easier to commit these short generic names to memory, but Linnaeus didn't at first apply the same system to the species name, which he continued to base on lengthy Latin descriptions. Introducing short names to distinguish species was a later development, inspired by a shorthand system developed by his students that allowed them to instantly

NAME THAT ORGANISM

Most of us are familiar with *Tyrannosaurus rex*, the name given by scientists to the terrifying meat-eating dinosaur that lived about 65 million years ago, but you may not be aware that every species of animal or plant is assigned a two-part Latin name, or binomial. The first word is the name of the genus to which related species belong, in this case *Tyrannosaurus*. The second distinguishes the individual species, in this case *rex*. The rules are complex, but essentially the correct name for a new species is the first binomial to be published in a reputable journal, usually by the scientist who discovered or first described it, and the species name must always be used with its generic name. When scientists come to publish their research findings about a new organism, awarding this unique name ensures that they avoid any ambiguity, and is an essential aspect of classification. Also, it helps biologists from different countries to communicate with one another, since the binomial crosses language boundaries, whereas the common name, if there is one, differs from one language to the next.

Scientists often name new species after respected colleagues or scientists from the past. John Gould, for example, identified the remains of a Christmas lunch, preserved by Darwin during his great voyage on the *Beagle* as belonging to a new species of bird from the genus *Rhea*. Although Darwin had no idea that he and the crew had eaten a new species, he methodically labelled and wrapped the inedible remains and returned them to England as part of his collection. Gould named the bird *Rhea darwinii* in honour of Darwin's contributions to the Zoological Society in Britain. Other factors might influence the choice of a name too. A new genus of Australian clam, for example, was recently discovered by a zoologist from The Natural History Museum, London who couldn't resist naming it *Rastafaria thiophila*, in honour of the characteristic dreadlock-like tubes that trail from the edge of its shell.

Rastafaria thiophila.

identify and classify specimens in the field. The fifth edition of *Genera Plantarum* (1754) is now internationally agreed to be the starting point of modern botanical nomenclature, as is volume one of the tenth edition of *Systema Naturae* (1758) for zoology.

THE BIRTH OF GEOLOGY

From the beginning of the 18th century, surveyors, explorers, and naturalists, including Linnaeus during his journey through Lapland, spent an increasing amount of time scaling mountains, studying rocks, and observing the effects of wind and water erosion. Despite deeply-held religious convictions, by the middle of the century it was impossible to ignore the mounting evidence that the surface of the Earth was undergoing a process of continuous change, and that this change had taken place since its origin. The rapidly expanding fossil collections gathered from stratified (layered) rock provided the clue to the ancient history of the Earth. In 1705, Robert Hooke compared fossilized and living wood under his micro-scope, providing powerful evidence to support

Steno's suggestion that fossils were the remains of once-living animals. The fossilization of organisms that appeared in many successive layers of rock must have been extremely slow, suggesting that change had occurred gradually over a long period of time.

Evidence for a continually changing world encouraged naturalists to revisit the Bible. Following Newton's lead, they started to argue that God was responsible for the original creation of the Earth, but natural forces, such as wind and water erosion, had been altering the surface of the Earth ever since. By the end of the 18th century, two theories vied with each other to explain the shape of mountains, valley plains, and rock formations. The German mineralogist Abraham Werner (1749–1817) provided the strongest arguments in favour of the theory of Neptunism, which asserted that rocks, and eventually mountains, were precipitated from a retreating ocean that had once covered the surface of the Earth. The Scottish geologist James Hutton (1726–97) argued in favour of the competing Plutonist theory that the surface had been shaped by great sections of rock, lifted from beneath the surface of the ocean by a series of volcanic explosions and violent movements in the Earth. Neither theory would triumph, but Werner and Hutton triggered a debate that marked the point at which geology became a separate field of study, alongside botany and zoology.

The Grand Canyon, Arizona, USA.

ACCUMULATING THE EVIDENCE FOR EVOLUTION

From human origins, some two and a half million years ago, curiosity has helped us to unravel the mysteries of nature. It has fed, protected, dressed and healed us, fired our imagination and thrilled us with discovery. Exploration, the search for new trading routes, wealth and political power, pushed our curiosity in new directions. Discovering new lands brought a flood of new animal, plant and mineral specimens back to Europe, and naturalists started to piece together the geography of the world. Inspired by Newton's success, they made two important advances during the 18th century: a convenient classification system for imposing order on a previously chaotic jumble of specimens, and recognition that the Earth was undergoing a process of continuous change. These developments led to a flurry of activity and an increasing accumulation of evidence that, along with changing political and cultural attitudes, eventually encouraged even the most devout members of society to question traditional assumptions. It would take another sixty years before Darwin and Wallace announced their theory of evolution by natural selection, but the foundations had been laid for the revolution that would finally transform natural history into the modern science we recognize today.

Temple to Nature

'He valued it at fourscore thousand;
and so would anybody who loves
hippopotamuses, sharks with one ear,
and spiders as big as geese!'

Horace Walpole, Trustee of the
Hans Sloane Collection, in a letter to
Sir Horace Mann, February 14, 1753

The towering façade of the Waterhouse building of The Natural History Museum, London. The basalt columns of Fingal's cave in Scotland inspired the design of the grand entrance.

It's impossible to ignore the building that houses The Natural History Museum, London. Unmistakably Victorian, the towering architecture dominates the landscape, a powerful reminder of Britain at the height of her imperial power. Even today, the walk through the great wooden doors at the main entrance can be a curiously humbling experience. Was this blue-and-buff coloured building, with its arched windows, towers and columns, once a cathedral? Look more closely at the elaborate carvings of animals, birds and fishes, and the paintings of plants that decorate the ceilings, and you'll realize that it can't have been, that it must have been created for the same purpose that it serves today. So if it wasn't originally designed to be a cathedral, why does it feel like a place of worship?

The simple answer is because this was the intention of the man who helped to design it — orthodox creationist Richard Owen, the brilliant 19th century comparative anatomist who became the Museum's first superintendent in 1881. Perhaps a more revealing question is why

LEFT: **Charles Darwin.**

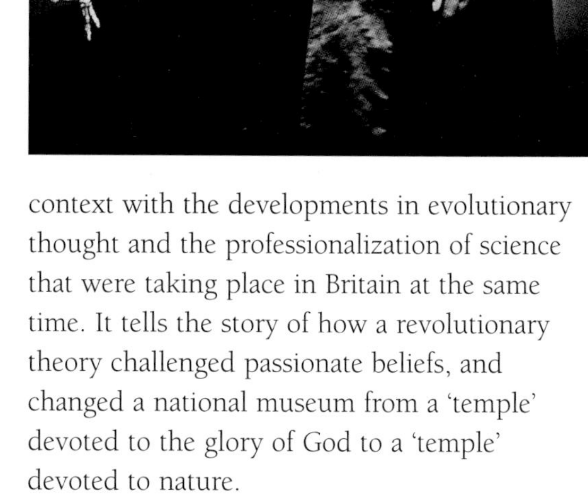

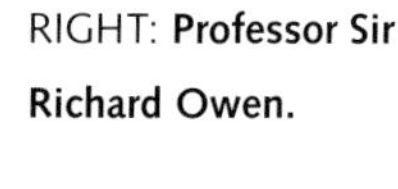

RIGHT: **Professor Sir Richard Owen.**

many 20th century visitors to the building fail to associate natural science with religion. To answer this, we must turn to the revolution that changed the study of natural history from amateur indulgence to a professional science over 100 years ago.

It is ironic that besides intending to provide a monument to God and a celebration of nature, Owen also wanted the new national museum to reflect adequately the success of the British Empire. As we will see later in this chapter, it was imperialism that gave Charles Darwin and Alfred Russel Wallace the opportunity to travel around the world, and imperialism that allowed them to formulate the theory that would revolutionize natural history and discredit the creationist beliefs so dear to Owen's heart.

This chapter looks at the events that led to the creation of The Natural History Museum in context with the developments in evolutionary thought and the professionalization of science that were taking place in Britain at the same time. It tells the story of how a revolutionary theory challenged passionate beliefs, and changed a national museum from a 'temple' devoted to the glory of God to a 'temple' devoted to nature.

A PRIVATE COLLECTION BECOMES PUBLIC PROPERTY

The Natural History Museum owes its beginnings to the independent 18th century collector Sir Hans Sloane. A wealthy physician and president of both the Royal Society and Royal College of Physicians, he acquired over 80,000 individual items, forming the single largest collection of any individual in Europe. This included a valuable herbarium, a huge

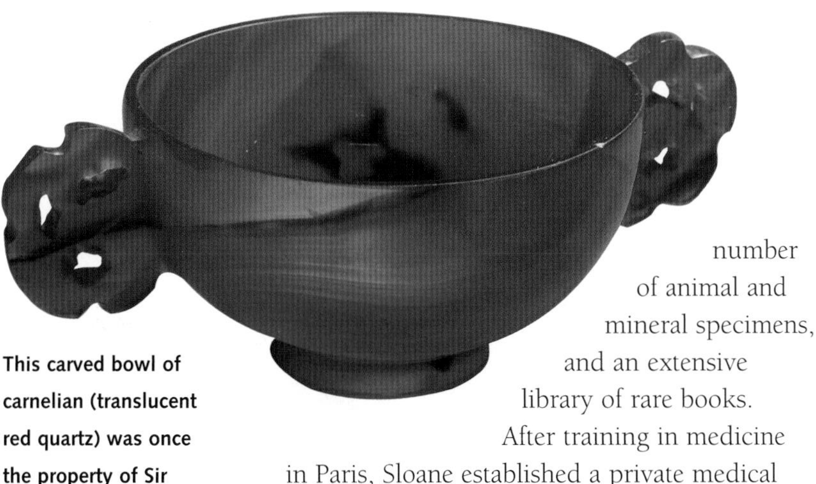

This carved bowl of carnelian (translucent red quartz) was once the property of Sir Hans Sloane.

number of animal and mineral specimens, and an extensive library of rare books.

After training in medicine in Paris, Sloane established a private medical practice in London. Then, in 1685, he accepted the post of physician to the governor of Jamaica, the Duke of Albemarle: it was during his travels in the West Indies between 1687 and 1689 that he gathered a large part of his collection. After returning to London, he lived in Bloomsbury until 1742, when he moved his collection of specimens and rare books to a private museum and library in Chelsea. He continued to add to his collection throughout his life and, with its many 'curiosities', it became one London's most popular attractions.

Sloane died in 1753, at the grand age of 92, leaving his collection in the hands of 51 trustees, all carefully selected friends or relatives. Corruption was widespread towards the end of Sloane's life, penetrating the highest levels of

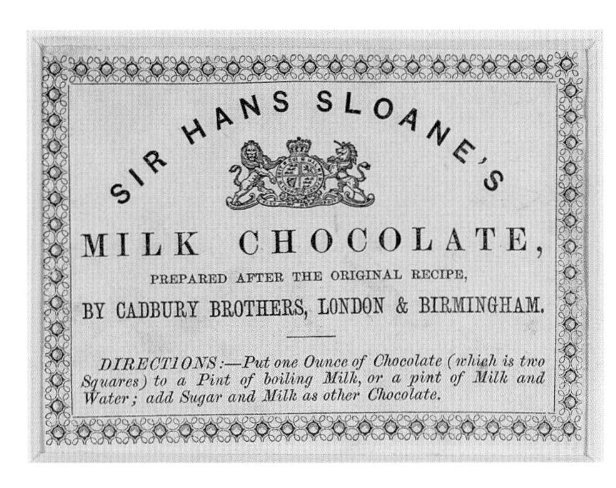

In the West Indies, Sloane observed the custom of brewing a dark, bitter drink by boiling the beans from a local tree. He found the taste 'nauseous', and added sugared milk to the recipe. He sold the recipe to a local apothecary in London, and eventually the Cadbury family acquired it.

Sir Hans Sloane, physician, naturalist and founder of the British Museum.

British society, and he was understandably cautious. Nominating 40 highly respected and influential Fellows of the Royal Society might seem over-zealous, but it was the best protection he could offer for a lifetime's work.

Sloane's caution paid off, and, three months after his death, an Act of Parliament agreed to purchase his Chelsea collection and the Harleian collection of manuscripts. They were to be stored, together with the Cottonian library, in a single location for the dual purpose of serious academic study and public enjoyment.

The token fee to be paid to his daughters that Sloane asked for in his will was a small price to pay for the most extensive collection in Europe, but King George II wasn't exactly eager to provide the funds, declaring 'I don't think there are twenty thousand pounds in the Treasury'. A national lottery was organized to

The Tartarian Lamb, dating from 1698, was formerly in the collections of Sir Hans Sloane. This plant was once thought to grow on a stalk but eat grass like a lamb.

produce the money instead. It shocked polite society, but achieved its aims, and in 1759 Montagu House in Bloomsbury, London, opened its doors to the public as the British Museum.

Early Years at the British Museum 1760–1799

James Empson continued to act as curator of Sloane's collections after the move to Bloomsbury, until he retired through ill health, while Gowan Knight was appointed the museum's first principal librarian, responsible for its collection of rare books. When Knight and Empson retired, their duties were combined, and Matthew Maty was appointed the new principal librarian, now responsible for the management of the library and all other collections at the British Museum. In 1763, the Swedish naturalist Daniel Carlsson Solander (1736–1782), an authority on the Linnaean classification method, became assistant librarian. He was one of Linnaeus's favourite students, and it was on the great botanist's advice that the 24-year-old first decided to travel to England. He soon set about the daunting but much needed task of systematically classifying and cataloguing the natural history collections.

Solander's reputation spread, and in 1764 the young Joseph Banks, who was destined to become president of the Royal Society, trustee of the British Museum and one of the greatest patrons of natural history in the 19th century,

Montagu House in Bloomsbury — the first home of the British Museum — in about 1800.

approached him for his advice on collecting plants. Starting what would become standard practice in the 19th century, Banks was planning to travel as a naturalist with Captain Cook on his first voyage around the world. The lure of faraway lands and a cornucopia of undiscovered plant and animal life were irresistible, and in 1768 Solander found himself aboard Cook's ship *Endeavour*, at the start of a journey that would last three years. Banks and Solander were accompanied by the young artist Sydney Parkinson, who provided a scientific record of their collections in a series of stunning coloured illustrations. Parkinson produced thirteen volumes of plant illustrations and three of animal illustrations, which included the first sketch by a European of a kangaroo. Tragically, though, he would never see his prodigious work published. Aged only 25, Parkinson died from malaria during the long return journey home.

One of Linnaeus's 'apostles' — the Swedish naturalist and physician, Daniel Carlsson Solander.

Respected scientist, energetic patron and enthusiastic sponsor, Sir Joseph Banks was responsible for much of the great scientific art of the late 18th and early 19th centuries.

This specimen of *Banksia serrata*, named in honour of Banks, was collected and painted during Captain Cook's first voyage.

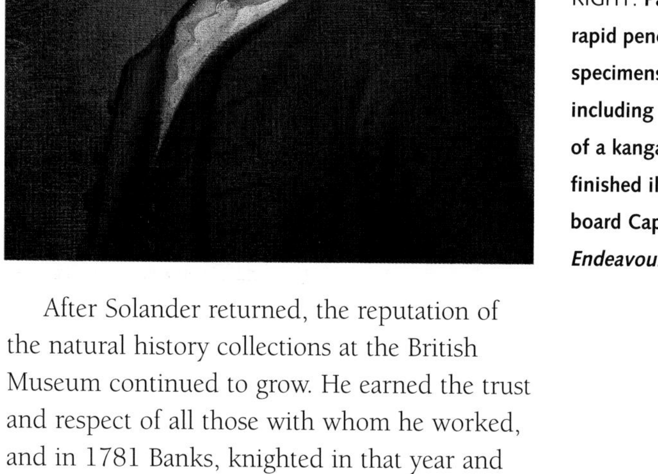

LEFT: **The gifted Scottish artist Sydney Parkinson.**

RIGHT: **Parkinson made rapid pencil drawings of specimens in the field, including this famous sketch of a kangaroo, to work into finished illustrations on board Captain Cook's ship *Endeavour*.**

After Solander returned, the reputation of the natural history collections at the British Museum continued to grow. He earned the trust and respect of all those with whom he worked, and in 1781 Banks, knighted in that year and an influential trustee of the British Museum, had no difficulty in persuading the Royal Society to place its natural history collection under the assistant librarian's able care. Then, in the following year, Solander died suddenly, at the age of 46. Neither his immediate successor Paul Henry Maty nor Edward Whitaker Gray, who was made keeper of the natural history collections in 1787, did much to enhance the legacy Solander left behind. In 1799, two outstanding private mineral collections, the

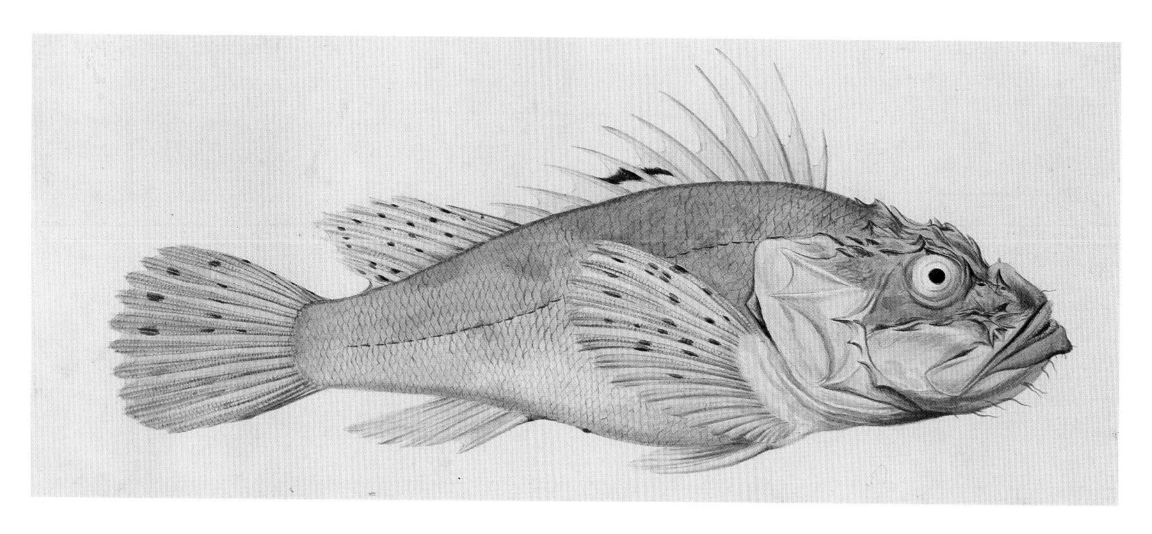

Illustration of a scorpion fish, by Sydney Parkinson.

Charles Hatchett Collection and the Mordaunt Cracherode Collection, were acquired by the museum. These acquisitions marked an otherwise unremarkable phase in its history, representing fine replacements for Sloane's more amateurish offerings.

Drowning Under a Deluge 1800–1849

A DETERIORATING COLLECTION

In 1806, the appointments of George Shaw as keeper of the new Department of Natural History and Modern Curiosities and the German naturalist Charles König as his assistant, heralded the beginning of a new phase in the history of the British Museum. Sadly, it also marked the point at which most of Sloane's deteriorating zoological specimens would be incinerated, as part of a new campaign to organize the chaos of specimens overflowing within the basements of the building. Only limited techniques for preserving specimens were available in the 17th century, and some were more successful than others. Fish, for example, were dried before the superior technique of storing them in alcohol was developed in the 1660s, and many of the animals collected by Sloane had either rotted away or become damaged beyond repair by the 1800s.

Even Banks, by now a principal trustee of the British Museum, had grown disheartened by the way so many specimens had degenerated, and donated a large part of his private zoological collection to the great Scottish physiologist and surgeon John Hunter (1728–93), to start a new museum attached to the Royal College of Surgeons. But Shaw, König and the new librarian, Joseph Planta, had

Key events in the development of natural history, 1750–1850.

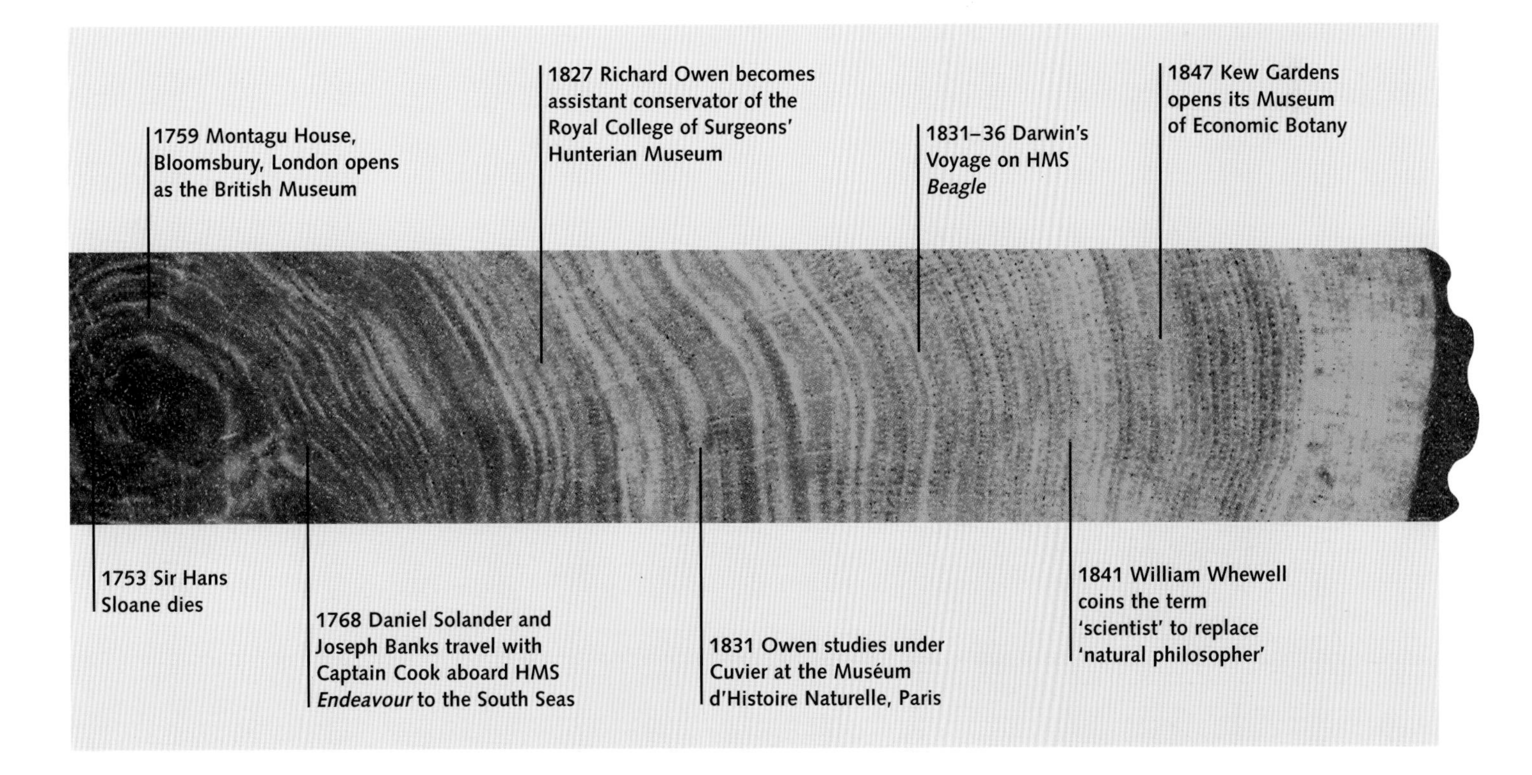

ambitious plans to mirror the success of France's flourishing Muséum d'Histoire Naturelle, and Banks gradually began to regain his old confidence in the naturalists at the British Museum.

Banks showed his approval of the revitalized department when he bequeathed his valuable botanical collection, including the plants he had gathered with Solander, to his curator, Robert Brown, on the understanding that he would eventually pass them on to the museum. Sir Joseph Banks died in 1820. A generous benefactor and President of the Royal Society for 40 years, his enthusiasm and substantial private collections had been a source of inspiration to naturalists throughout his life.

Despite König's best intentions after he was made keeper in 1813, the British Museum remained under-staffed, short on space and a long way from achieving the status of the French natural history museum. A steady stream of specimens from expert naturalists, land surveyors and colonial enthusiasts continued to arrive at Montagu House from the far corners of the globe. The museum's experts struggled under the deluge of unidentified specimens, and according to radicals such as Robert Grant, professor of natural history at the University College of London, the aristocratic trustees were doing little to help. In 1835, unable to ignore growing concerns about corruption and inefficiency, the House of Commons set up a select committee to look into the conditions and management of the British Museum.

Dr John Edward Gray, keeper of the Zoological Department of the British Museum from 1840 to 1874.

The committee set up separate interviews with staff at the museum, and the evidence provided by the zoologist and botanist John Edward Gray (Edward Whitaker Gray's great-nephew) and the librarian, Antonio Panizzi, was to prove particularly influential in shaping the recommendations of their report. Using his first-hand knowledge of the Muséum d'Histoire Naturelle, Gray argued that the mineralogy and zoology collections should no longer be kept together as a single department. He also suggested that more money should be made available for research, and that if the experts were to make real progress, they would need to replace the cleaners and warders who currently supported them with properly trained assistants.

Although they held conflicting views regarding the relative importance of books and natural history, both Gray and Panizzi believed in creating a museum that served the interests of all, no matter what their financial or social station. Everyone should benefit from the nation's collections and have access to the advice of the museum's experts.

Concerned primarily with his precious books, Panizzi suggested that the problem of space would be relieved if the natural history collections were moved to a separate location. The committee ignored this costly idea, but did divide the natural history department into three sections: Mineralogy and Geology; Botany; and Zoology. Gray, who had been only a temporary member of staff, was appointed assistant to the keeper of zoology and, eventually, became keeper in 1840.

Loyd Grossman

CHAIRMAN OF THE CAMPAIGN FOR MUSEUMS

Any visit to The Natural History Museum, no matter how brief, must start with the building itself. I recall that Nikolaus Pevsner in *The Buildings of England* felt that this Romanesque German concoction which so dominates the museum quarter of South Kensington was rather dull. But the sight of it fills me — and I suspect many others — with joy.

I am a great fan of Alfred Waterhouse, the architect who struggled for so many years with this building, and I highly recommend Mark Girouard's outstanding book *Alfred Waterhouse and The Natural History Museum* to any visitor to South Kensington. I can add nothing to the understanding of the building, other than my own appreciation of it as possibly the greatest example in the world of the museum as a palace of knowledge. As you study the building itself, its marvellously complex decoration of species living and extinct, its intricate spaces and cathedral-like Central Hall, you will be made aware, as Waterhouse intended, of the majesty, complexity and fascination of the world around us. But the building is, of course, just the beginning. Dippy, the 26-metre-long *Diplodocus*, may get top billing in the Central Hall, but I also like to look at the statue of Richard Owen on the main staircase's first landing. As founder of the Museum — memorably described by a Victorian MP as 'so foolish, crazy and extravagant a scheme' — Owen pioneered the popular appreciation of natural history. My favourite part of 'his' Museum is a tribute to another 19th-century pioneer, Mary Anning, 'the fossil woman of Lyme Regis', who began her career by discovering a fossilized ichthyosaur when she was just 11 years old. The fossil marine reptiles displayed on the walls of Waterhouse Way represent a tribute to her hard work and curiosity, and are some of the most enthralling specimens in the Museum.

My children are devoted to the Earth Galleries. Of course, they love riding the giant escalator and being shaken by the evocation of the Kobe earth-quake, but they are also delighted and inspired by the Moon rock and the 25-million-year time trail.

The opening of the new Jerwood Gallery adds an unmissable new attraction to the Museum, and a chance to display some of the finest works from the Museum's collection of half-a-million drawings, paintings and prints. The Jerwood Galleries are also part of a process of revealing more of what one might call the hidden Museum, because of course visitors can only sense a small portion of what this institution is about. Visitors in the future will, I hope, get a greater appreciation of the world-class research, the million-volume library, and the commitment to lifelong learning that will make even the quickest visit the beginning of many more.

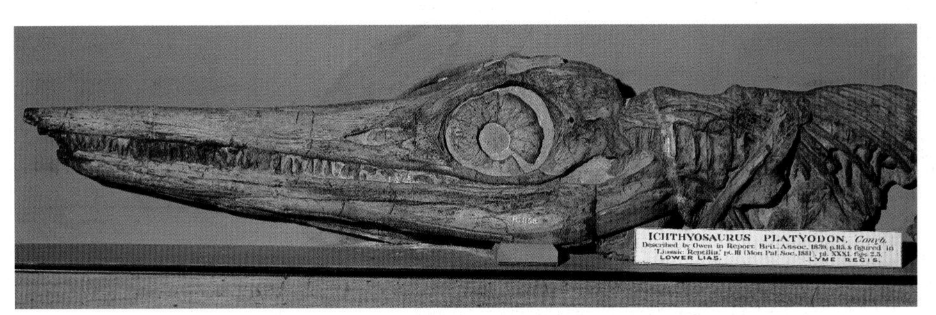

This icthyosaur skull, now called *Temnodontosaurus platyodon*, was collected by Mary Anning.

LAGGING BEHIND EUROPE

The nation's natural history collections remained within the British Museum for the next 40 years. From the museum's opening in 1759, natural history had shared storage and exhibition facilities with the arts. By the first half of the 19th century the fate of its collections was a reflection of the British Government's *laissez-faire* attitude towards a troublesome area of science. The expanding collections were regarded more as a treasury of objects than a resource for learning and study. Limited funds and an increasing shortage of space restricted public access to the collections, made caring for them extremely difficult, and offered little by way of incentive for research.

The contrast with the situation during the same period in France couldn't have been more startling. By the early 1800s, the revolutionary government's newly created Muséum d'Histoire Naturelle in Paris had established the French as world leaders in the empirical study of nature. In addition to its more traditional activities, the French museum pioneered a unique teaching and research programme that rapidly became the envy of the world. In a bid to expand education and research opportunities for aspiring naturalists, the Paris museum funded twelve professorships, with dedicated teams of skilled support staff, and attracted a constant stream of international visitors keen to learn from this new body of professional scientists.

Casting a wary eye over the activities of their neighbours, the British ruling class associated this move to professionalize science with revolution and anarchy. Knowledge could be dangerous in the wrong hands and, far from persuading the establishment to follow suit, the example set by France only convinced them to leave the study of natural history to the trustworthy few — members of the clergy,

An early illustration of the Muséum d'Histoire Naturelle in Paris.

gentlemen and dedicated naturalists lucky enough to secure a patron.

CHURCH INFLUENCE

Living in the modern secular world, it is difficult for many of us to appreciate the importance of religious beliefs and the influence of the Church in Britain during the early 19th century. The Church encouraged religious dependency by promoting the belief that God acted through the clergy to control the spiritual fate of humankind on Earth, and the ruling aristocracy was happy to support the Church because it believed dependency would prevent civil unrest. This was a society in which educational and political opportunities were reserved for members of the Anglican Church, and any sign of dissent against orthodox views was deemed not only an attack on religion, but also politically subversive and a threat to the establishment.

Natural history continued to retain close links to theology, with much of the work in the early 19th century being devoted to classification and description. Although most naturalists continued to believe that God had created every individual species of plant and animal, and that the world was static and unchanging, a small number of radical freethinkers were beginning to speculate that living things may have originated and evolved as a result of entirely natural processes. In 1809, the great French naturalist Jean-Baptiste Lamarck (1744–1829), professor of invertebrate zoology at the Muséum d'Histoire Naturelle, published the most comprehensive of these theories in his *Philosophie Zoologique*, which proposed that species adapted and evolved by a process called 'the inheritance of acquired characteristics'.

These theories of evolution challenged orthodox Anglican theology, and horrified the British establishment. The idea that species evolved as a result of random natural processes raised questions over the existence of a divine power, and challenged the legitimacy of the

LAMARCK AND THE INHERITANCE OF ACQUIRED CHARACTERISTICS

The French 'father of evolution', Jean-Baptiste Pierre Antoine de Monet Chevalier de Lamarck.

Lamarck believed that life had started as a result of a chemical reaction induced by a spark of electricity, and that primitive forms had gradually evolved to produce more complex organisms.
He argued that plants and animals had evolved in separate, but parallel, sequence, with humans representing the most complex form of all. From his studies in taxonomy, Lamarck knew that species didn't fall into a linear pattern of complexity. This, he argued, was because living organisms were undergoing a process of continuous transmutation or change, adapting to fluctuating conditions in their environment. Lamarck believed that species acquired new characteristics when they encountered a different environment, and that these characteristics were passed on to their offspring to enable them to adapt and survive. He argued that giraffes, for example, had gradually developed long necks as a result of successive generations stretching to eat leaves on trees. Exercising the neck encouraged it to grow longer, and each generation of adult passed on these incremental increases to their offspring. One of the major weaknesses in Lamarck's theory lay in the fact that he refused to believe that nature would allow species to become extinct, despite new fossil evidence to the contrary. Biologists would have to wait another century to show that there was no mechanism by which acquired characteristics could be imprinted on genes and passed on to subsequent generations.

Darwin, in about 1857, just before he published *On the Origin of Species*.

Church. Political radicals used evolution to support their argument for self-improvement and the democratization of society, and evolution, like the professionalization of science, became associated with subversion and revolution.

The two political parties of the day, the ruling aristocratic Tories and the more liberal-minded Whigs, closed ranks with the Church to condemn these subversive ideas and maintain the status quo. Lamarck's theory was condemned as abhorrent, with orthodox naturalists such as Edward Whitaker Gray, keeper at the British Museum until 1806, seething at the 'abominable trash vomited forth by Lamarck and his disciples ... who have rashly, and almost blasphemously, imputed a period of comparative imbecility to Omnipotence'. Nowhere else in Europe did natural history become so politicized. It was in this atmosphere that Richard Owen began his campaign to establish a national museum dedicated to natural history, and Darwin started to formulate the theory destined to transform the study of natural history into a science.

A VISIT FROM A RETURNING HERO

In 1836, while Gray and Panizzi were making their report to the government select committee, Charles Robert Darwin, freshly returned from his epic journey on the *Beagle*, paid a visit to the British Museum. It was the first stop in an exhausting survey of the London museums to find a suitable home for his impressive collection. The quintessential Victorian naturalist, Darwin had been a tireless collector, methodically gathering examples of every fossilized and living organism that he could find. He had carefully dissected, preserved and catalogued his specimens in preparation for expert opinion, and he wasn't going to hand his treasures over without a guarantee of swift and accurate attention.

A LIBERAL EDUCATION

But what led Darwin, a gentlemen of privilege, to ponder over ideas that challenged orthodox beliefs underpinning the society that had given his family so many advantages? Born in 1809, the grandson of the freethinking evolutionist Dr Erasmus Darwin and the famous pottery industrialist Josiah Wedgwood, Charles was no stranger to liberal ideology. Encouraged to pursue a career in medicine, he continued his liberal education at Edinburgh University where his father, Dr Robert Darwin, believed he would receive a more rounded medical education than at the conservative, Church-dominated universities of Oxford and Cambridge.

Darwin was so traumatized by an operation that he witnessed performed on a child without anaesthetic that he gave up his studies without completing the course, and filled his time instead with a growing fascination in nature. He joined the Plinian Society, which provided a forum for debate on the latest developments in science. The debates were often political, highly charged exchanges between radical students arguing against supernatural explanations of nature and more orthodox members who upheld conventional religious beliefs.

POWERFUL INFLUENCES

It was during a meeting of the Plinian Society that Darwin first met Robert Edmund Grant, a trained doctor, who had given up his career to study marine zoology. Fiercely anti-Christian, Grant was one of the few naturalists of his time to openly support Jean-Baptiste Lamarck's theory of evolution. Although Darwin was already familiar with his grandfather Erasmus's theory of evolving species, as a result of reading his medical treatise *Zoonomia*, Grant's irreverent defence of transmutation must have influenced him. Darwin also learnt important practical skills in dissection and classification from his unconventional companion.

The marine zoologist Dr Robert Edmund Grant, an important early influence on Darwin.

Having abandoned medicine, Darwin enrolled as an undergraduate at Christ's College, Cambridge, to pursue a career in the Church. It was here that Darwin first met John Stevens Henslow, the professor of mineralogy and botany who was to become one of the greatest influences on his life. Henslow hosted evening gatherings for expert naturalists and students that were very different from the controversial Plinian meetings in Edinburgh. Only orthodox views were voiced, and Darwin settled down to learn all he could from the great Cambridge intellectuals investigating nature within the acceptable bounds of the Church.

DARWIN'S GREAT VOYAGE

It was Henslow who recommended the book that first inspired Darwin's ambition to travel. The vivid descriptions of exotic jungles and volcanic landscapes recorded by the German naturalist and explorer Alexander von Humboldt (1769–1859) during his expedition to South America had the young naturalist brimming with excitement. When the Admiralty approached Henslow, looking for a suitable gentleman to act as companion to Robert Fitzroy (1805–65), meteorologist and captain of HMS *Beagle* on a coastal survey of South America, it seemed only natural that he should put Darwin's name forward for the task.

There was plenty of time for reading during the long hours at sea. Von Humboldt encouraged Darwin to look at the relationship between animals, plants and their environment as a unified whole rather than unrelated fields of study, and a letter from Darwin's father set him thinking about the ideas of the political economist Reverend Thomas Robert Malthus (1766–1834). The early 19th century was a time of both extreme wealth and extreme poverty in Britain. The Government had just introduced harsh new laws to deal with the escalating problems, reviving an article written by Malthus in 1798 entitled 'An Essay on the Principle of Population' to justify its actions.

Malthus had a gloomy vision of the world. He argued that the growth of human population would eventually outstrip space and food supplies, and that without cruel but necessary controls like disease, famine and war, human population would double in only 25 years. Darwin realized that the same principles could be applied to plant and animal populations in the wild, and this gave him the idea for his theory of natural selection.

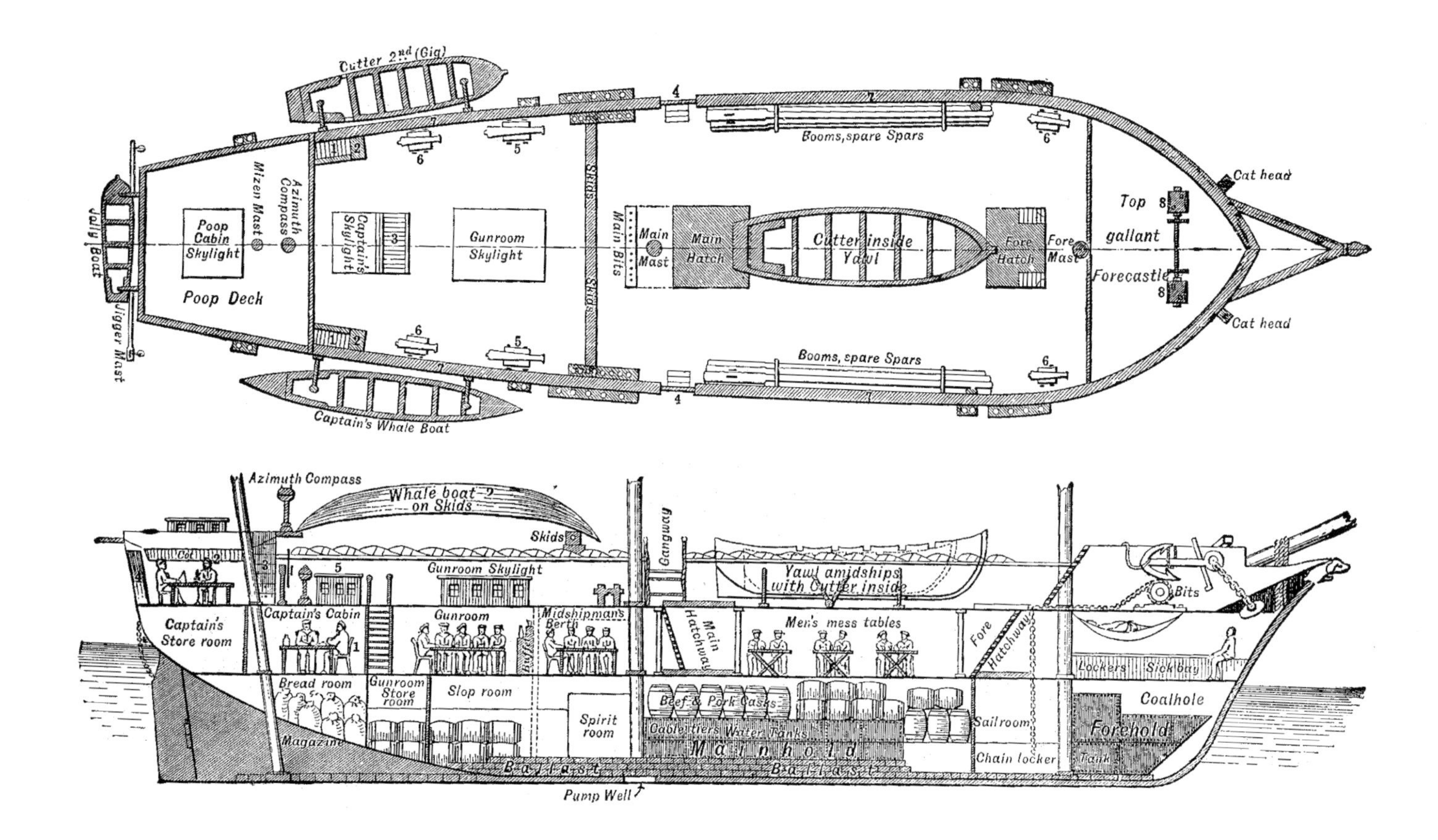

Having spent several days in the field before his trip on the *Beagle* honing his geological skills, under the guidance of the great Cambridge geologist Reverend Adam Sedgwick (1785–1873), Darwin was eager to read on the long voyage a best-selling book, *Principles of Geology*, written by another eminent geologist, Charles Lyell (1797–1875). Lyell had resurrected the 50-year-old theory of uniformitarianism, proposed by the Scottish geologist James Hutton (1726–97), and embellished it with his own observations. An orthodox Christian, Lyell believed that after the Creation, the Earth had been subject to gradual change through the action of present-day processes such as earthquakes and erosion, and they occurred over a period of time far longer than anything that had previously been accepted by geologists.

Darwin's geological observations in South America soon dispelled any doubts that the young explorer might have felt on first reading Lyell's vision of gradual change. While staying in Valdivia, Chile, for example, he witnessed tremors from the 1835 earthquake that devastated the city of Concepcion. Visiting the site of the tragedy soon after, he saw that the quake had forced mussel beds above the high tide mark and pushed great slabs of rock upwards on the beach. The violent movement had caused the land to rise by several feet. For Darwin, the earthquake confirmed that Lyell was right — mountain ranges were created not by a single massive disturbance, but by the succession of many small upheavals over great expanses of time. As Darwin commented, 'the Earth is a mere crust over a fluid melted mass of rock'.

Conditions on the *Beagle*, which was only 27.5 m long and 7.3 m wide, were extremely cramped. For five years, Darwin's home was the corner of the poop cabin, where he slept in a hammock strung over the Captain's chart table.

THE THEORY OF EVOLUTION BY NATURAL SELECTION

Darwin's ideas about evolution and the origin of species were inspired by extensive observations of natural phenomena made during his long voyage on the *Beagle*. He believed that far from being created at the beginning of time, or at separate punctuated intervals, the great diversity of life on Earth had emerged as a result of species changing or 'adapting' in response to a particular way of life.

Fossil evidence indicated that species that had lived in the distant past were not the same as the ones that lived in the 19th century, suggesting that changes had occurred extremely slowly over successive generations. Darwin predicted the discovery of fossilized species, such as the primitive bird *Archaeopteryx*, found in 1861 in Bavaria, that represented intermediary stages along evolutionary pathways. In the case of *Archaeopteryx*, for example, the combination of reptilian and bird-like features — feathers, teeth, a long bony tail and clawed fingers on each wing — provided persuasive evidence that birds had evolved from reptilian ancestors. Darwin believed that evolution occurred by a process of 'natural selection'.

He noticed that individual members within a population vary, and that this variation tends to be transmitted to the offspring of subsequent generations. He also noted that environmental resources are limited, and individuals are forced to compete with one another to survive. Nature selects individuals that display the most favourable characteristics, and they survive to generate more offspring with the same advantageous adaptations. Favourable characteristics build up over successive generations, and species gradually change or evolve. In the case of a variable population of giraffes, for instance, some individuals may have been taller than others. The shorter giraffes may have been unable to reach as much food, their resulting poor nutrition making them less likely to survive and breed. The taller giraffes with better access to food would be more likely to survive and reproduce, to create a new generation of taller individuals.

ACCUMULATING EVIDENCE

Darwin declared himself Lyell's 'zealous disciple' and spent much of his time during the voyage searching for evidence to support the geologist's theory. He made a number of important observations, and began to formulate his own theories about South American geology, volcanic islands and the formation of coral reefs. These were all later to become important to his evolutionary ideas, because they provided evidence of the Earth as a continously evolving planet. With the help of John Henslow, Darwin's observations were published while he was still at sea, and he was delighted to find on his return to Britain that he had already become something of a celebrity amongst the geological elite.

The geologist Sir Charles Lyell, one of Darwin's closest friends.

Now a rising star, Darwin was prepared to use what influence he could to have his collection viewed as a matter of urgency. But neither the British Museum, the subject of a government investigation and awash with unidentified specimens, nor the Zoological

Society's new museum, already crammed to capacity, offered an appealing option. Instead, Darwin turned to his father for financial support so that he could manage the collection himself.

DIVIDING THE SPOILS

The collection was divided into parts. Henslow had already agreed to work on Darwin's botanical specimens, most of the reptiles went to London's King's College and the geologist-clergyman Reverend William Buckland (1784–1856), Reader in Mineralogy at Oxford University, offered to take the iguanas from the Galapagos Islands. Darwin commissioned private experts, too. The expert shell trader George Sowerby accepted the fossil shells, while the birds went to John Gould (1804–81), the great ornithologist, artist and publisher, famous for his popular illustrated bird books. It was Gould, an expert who had risen from the humble ranks of a taxidermist at the Zoological Society, who revealed that the extraordinary variety of 'blackbirds' and 'gross beaks' collected by Darwin on the Galapagos Islands were all in fact closely related species of finches. The birds displayed widely differing beaks, each apparently perfectly adapted to exploit a specific type of food. Darwin believed that the finches had probably all evolved from a single species of finch that had somehow found its way to the islands from mainland South America.

Charles Lyell was delighted at the prospect of meeting Darwin in person for the first time, and looked forward to helping his young protégé settle into London society. It was on the evening of their first meeting that Lyell introduced Darwin to a tall man with penetrating eyes, the newly appointed Professor of Comparative Anatomy at the Hunterian Museum in London and future superintendent of The Natural History Museum. His name was Richard Owen.

LEFT and RIGHT: **Finches from the Galapagos Islands by John Gould, from *Zoology of the Voyage of the Beagle* by Charles Darwin, published 1832–1836.**

The young Richard Owen. Thomas Henry Huxley said of him 'The truth is he is the superior of most and he does not conceal that he knows it'.

Owen was already familiar with Darwin's sensational discovery of fossils belonging to giant, extinct South American animals. Several years earlier, an impressive skull, as well as teeth, leg and jawbones, had been posted to Henslow in Cambridge, where they had become the source of great excitement. An ambitious man, Owen had gradually monopolized the Zoological Society's supply of dead animals for dissection, developing an unrivalled expertise in the anatomy of exotic wildlife. Ever alert for new trophies, he knew that Darwin's latest giant fossil discoveries held the promise of even greater accolades. Darwin, for his part, was delighted to accept Owen's swift offer of help with this last but most exciting aspect of the *Beagle* haul.

THE MAN WHO NAMED THE DINOSAURS

Owen was born in Lancaster in 1804, the son of a West Indies merchant and a French Huguenot refugee. Having studied medicine at Edinburgh University, he started his career assisting the highly respected surgeon John Abernethy (1764–1831) at St. Bartholomew's Hospital in London. After qualifying as a surgeon in 1826, he set up in private practice. A year later, he accepted the post of assistant conservator of the Hunterian Museum, attached to the Royal College of Surgeons. In 1830, the famous French comparative anatomist Georges Cuvier (1769–1832) visited the Hunterian Museum, and was so impressed with the scientific abilities of the French-speaking Owen that he invited him to study comparative anatomy and palaeontology in Paris.

Owen spent a short time at the Muséum d'Histoire Naturelle in 1831, and mastered Cuvier's method of predicting the appearance and the behaviour of ancient animals from the smallest fragments of fossilized remains, such as a single tooth or a well preserved piece of bone. Cuvier's fossil specimens were taken from stratified rock in the Paris basin. Each individual layer of rock, representing a specific period of time in the history of the Earth, contained its own unique population of fossils. Cuvier believed that there was no connection between the different populations, and that natural catastrophes, floods or earthquakes had caused the extinction of

The great French anatomist and palaeontologist Georges Cuvier.

successive populations. Owen, a devout Anglican, believed the catastrophe theory agreed perfectly with evidence from the Bible, and used Cuvier's arguments in his skirmishes with radical scientists who preached transmutation and evolution.

CONTEMPT FOR OPPONENTS

What would Owen's reaction have been on first meeting Darwin, had he known the young explorer was harbouring secret thoughts of evolution? A staunch conservative and champion of the establishment, Owen launched an aggressive campaign to discredit Darwin's old companion from Edinburgh, the outspoken radical Robert Grant. This gave Darwin ample cause for remaining silent about evolution. Grant's expertise in comparative anatomy had led to his appointment as Professor of Natural History at the new University College of London, placing him in the spotlight of a small but highly active scientific community. London wasn't big enough for two leading experts in comparative anatomy, and Owen was tireless in his efforts to see Grant undermined.

Incensed by the Scottish radical's accusations of amateurish mismanagement at the Royal College of Surgeons and the British Museum, Owen voted his rival out of a position at the Zoological Society. In December 1838, when Darwin was secretary of the Geological Society, he witnessed at first-hand Owen's contempt of Grant and his Lamarckian theories. During a talk given by the unsuspecting Grant to a meeting of the Geological Society, Owen and the Reverend Buckland launched a pre-planned attack, to discredit his latest ideas. It was a chilling display, and the fear of a similar fate, professional humiliation and ridicule, was certainly an important factor in Darwin's decision to keep his ideas about evolution to himself.

INTRODUCING THE DINOSAURS

In 1841, at a meeting of the British Association of the Advancement of Science, Owen achieved something of a celebrity status when he presented fossilized evidence of a new order of reptiles that he called Dinosauria, from the Greek words meaning 'terrible lizard'. His startling descriptions of fearsome flesh-eaters and giant, lumbering plant-eaters immediately

Benjamin Waterhouse Hawkins drawings of *Megalosaurus*, *Iguanodon* and *Hylaeosaurus* for Crystal Palace, London.

captured the public imagination, and triggered a fascination with the dinosaurs, as they became more popularly known, that continues to this day. The fact that Owen based his description of this new order of reptiles on the sketchy evidence provided by only three species known at that time, 60 years before the discovery of dinosaurs such as *Tyrannosaurus rex*, bears testament to his remarkable anatomical expertise and creative brilliance.

By 1840, Owen's favour had spread to the Royal family. Queen Victoria granted him a private residence, Sheen Lodge in Richmond Park, and a civil list pension to supplement his income from the Hunterian Museum. By 1851, Owen was teaching natural history to the Royal princes and princesses, and working together with Prince Albert on preparations for the Great Exhibition. In 1854, at Prince Albert's request, he also supervised Benjamin Waterhouse Hawkins' life-sized reconstructions of *Megalosaurus*, *Iguanodon* and *Hylaeosaurus*, which were to appear on permanent display in the grounds of Crystal Palace after their removal from Hyde Park to Sydenham. Bolstered by ruthless ambition and thinly disguised social climbing, there seemed to be no limit to Owen's

OWEN AND THE DINOSAURS

The model dinosaurs on display at the Crystal Palace were an instant success but, with only limited evidence at his disposal, Owen's reconstructions of *Megalosaurus* and *Iguanodon* were not entirely accurate. Owen depicted them walking on four legs, but the discovery of complete skeletal remains of *Iguanodon* in 1877, in Belgium, revealed that these animals had huge hind legs, and a long fleshy tail — features suggesting that they were bipedal part of the time. Gideon Mantell, in his initial reconstruction of *Iguanodon*, assumed that the fossilized thumb was in fact part of the dinosaur's skull, and placed it, rhinoceros-fashion, on its nose. Both examples illustrate the difficulty, for even the most brilliant comparative anatomist, of making assumptions about extinct animals from a few fragments of bone, with only the skeletons of contemporary animals for comparison.

This cartoon depicts a famous dinner party at Crystal Palace in 1853, when twelve distinguished guests ate their meal inside the half-completed model *Iguanodon*.

influence and achievements. Yet it was precisely these less attractive aspects of Owen's personality that often supported him through his 20-year campaign to create a national museum of natural history that in his own words was 'worthy' of the great British Empire.

A New Appointment 1850–1859

By 1854, thanks to 14 years of enterprising management by John Edward Gray, keeper of the Department of Zoology, the British Museum had developed the largest zoological collection in Europe. His achievement was all the greater because of the apalling conditions under which his staff were forced to work. Toiling away in the airless basement, squashed between endless jars of specimens, researchers found the light so poor that when the fog set in, they could barely see their hands, let alone the animals they needed to study in minute detail. Gray's department had gained such an impressive reputation that the council of the Zoological Society could no longer see the need for a second zoological museum in London, and in 1855 it sold its collections to the British Museum. Although the botany, mineralogy and geology branches weren't publishing as many papers as the zoology section, their collections continued to expand at an alarming rate. Space limitations at the museum were compounded by enthusiastic archaeologists stripping foreign lands of their ancient assets.

The acquisition by the English archaeologist and politician Sir Austen Henry Layard (1817–94) of two enormous Assyrian sculptures in 1848 and 1849 raised the question once more of splitting the collections and moving part of them to a separate location. The summary of a second select committee investigation into the British Museum made no mention of new facilities for the natural history collections, but it did recognize that they were equal, if not superior, to any in the world.

Easter Monday crowds flock into the zoological gallery of the British Museum in the 1850s.

OWEN TAKES OVER

In 1856, Owen accepted the new position of superintendent of the Natural History Departments at the British Museum. It was not a popular choice for the keepers, particularly Gray, who became the object of Owen's professional jealousy, but it was an inspired appointment in terms of the fate of the natural history collections.

In 1858, Owen was elected president of The British Association for the Advancement of Science, and during a presidential speech that year launched an initiative to have the natural history collections housed separately in a new national museum. Owen stressed that the British Museum was home to the finest collection, but to 'adequately describe, classify and display' its zoological specimens he needed more space. Appealing to British patriotism, he argued that 'Never was there so much energy and intelligence displayed in the capture and transmission of exotic animals by the enterprising traveller in unknown lands and by the hardy settler in remote colonies, as by those who start from their native shores of Britain. Foreign Naturalists consequently visit England anticipating to find in her capital and in her National Museum the richest and most varied materials for their comparisons and deductions.'

OWEN'S HALL OF WONDERS

A passionate creationist, Owen believed, like Cuvier, that the main groups of the animal kingdom consisted of individuals that exhibited variations around a basic ground plan or type. This he planned to illustrate by means of an 'Index Museum' in the main exhibition hall of the new natural history museum. This museum within a museum would be a place of wonders, including example specimens from all the major groups. Only a magnificent, spacious building could do justice to such an ambitious project, and it became one of the main motivations in his long campaign for a new national museum. As Owen observed, 'The very fact that the whales being the largest animals upon Earth, is that which makes it more imperative to illustrate the fact and gratify the natural interest of the public by the adequate and convenient exhibition of their skeletons'.

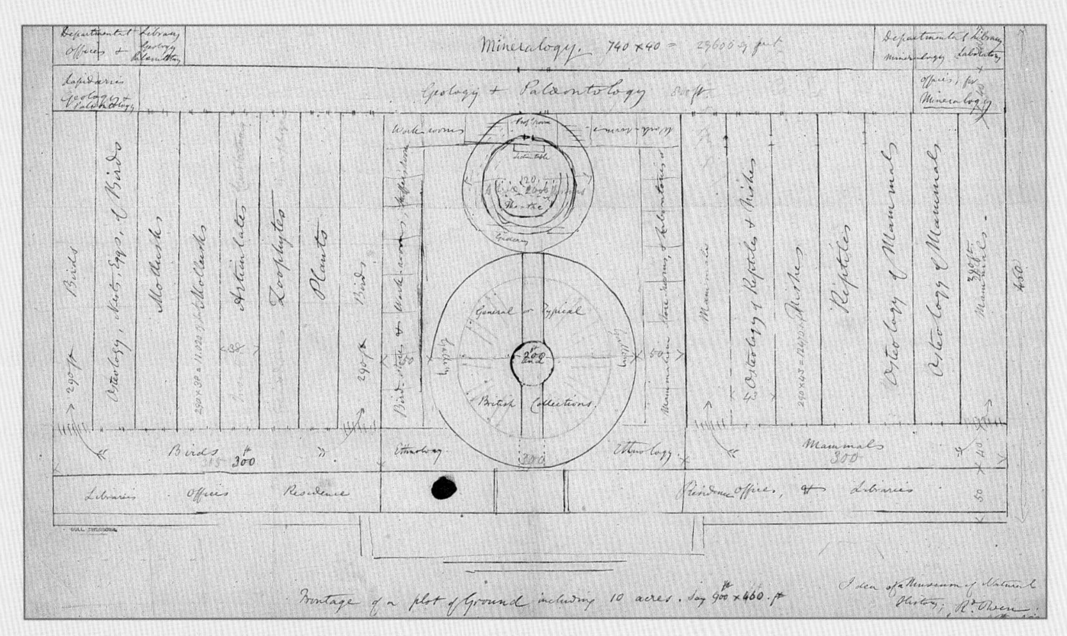

Owen's plan for an Index Museum, 1879.

Owen had a vision of creating a building that would celebrate the natural world and illustrate the scale of British overseas conquests through its wide-ranging display of exotic species. It would be a testimony to the museum's worth and a monument to God, nature and of course to his own achievements. According to the plan he drew up and presented to the Trustees in 1859, he calculated that he needed 4.5 ha for the single-storey building that would house the new museum.

The end of the Crimean war in 1856 had left little in the Treasury for costly changes to the British Museum, and 19 years of debate followed, with Owen calling on all his powers of influence and persuasion to see his dream become a reality. The famous Liberal politician William Gladstone (1809–98) became an important ally after another government select committee rejected Owen's plans in 1859. Determined to recruit Gladstone to his cause, Owen had gained the support of the rapidly overwhelmed politician, by taking him on a tour that included every inch of the dimly lit basements, bulging with the specimens he was unable to display.

SECRET NOTEBOOKS

While Owen was fighting his battles for a new building, Darwin was working away on his private theory. Although key ideas had occurred to him during his voyage on the *Beagle*, he didn't begin to formalize his theory until three years after his return to London. The notebook he opened in 1839 became the first in a long series recording 20 years of methodical research in minute detail. Darwin, ever the collector, hoarded everything — specimens, facts, every letter that he had received, every note and scrap of paper that he had written — leaving behind a unique record of his triumphs and setbacks in a relentless pursuit of truth.

His notebooks were a secret, and would remain firmly locked in his study at Down House, the family home in Kent. He had many reasons for this wariness. Professional ridicule was a worry, but equally important was the fashionable fear that any theory of evolution would provide ammunition for radicals looking for arguments to revolutionize society. Darwin was a compassionate man, who welcomed reforms that freed slaves and improved educational opportunities, but he saw no advantage in toppling the establishment. He was a gentleman, who relied on private investments in the railways and the many other lucrative opportunities to be had in an industrializing Britain. Why upset the system that supported him in his choice of careers and put food in the mouths of his rapidly expanding family?

Besides his political misgivings, Darwin had professional reasons for keeping his theory under wraps. Although he had already gathered some of the evidence for his theory during the five years he spent abroad, he needed more. He wanted to produce irrefutable proof, in part to protect himself from the accusations of quackery levelled at evolutionists like Grant, but also to convince himself of ideas apparently so at odds with his privileged place in society.

PLAGUED BY ILLNESS

Darwin worked on his theory of evolution at great personal cost to his health and his nerves. He conducted years of painstaking research, dissecting and classifying every known species of living and fossilized barnacle, breeding pigeons and orchids to test his ideas of artificial selection, and studying insectivorous plants and the habits of earthworms.

For the next 20 years of this great work, indeed for the rest of his life, concentrated periods of methodical observation were

A MEMBER OF THE PIGEON FANCIER'S ASSOCIATION

Darwin's tireless pursuit of evidence to support his hypotheses found him often investigating topics outside the conventional bounds of natural history. So, while his contemporaries concentrated on studying species in the wild, Darwin turned his attention towards the breeding of domesticated animals, such as pigeons. Darwin believed that pigeon fanciers mimicked nature by artificially 'selecting' desirable features — the shape of the feathers on the crown of the bird, for example — and breeding from the best stock to encourage new trends in variation amongst their offspring. Individuals without the desired characteristics were discarded, just as poorly adapted species perished naturally in the wild. Darwin believed that the only way to pick up the tricks of the pigeon breeding trade was to join numerous working men's clubs across London, and learn from the expert laymen at first hand. He was willing to attend even the lowliest of venues, including The Borough Club in South London, to learn from the experts, no matter how strange they seemed, writing 'I am hand & glove with Fanciers, Spital-field weavers & all sorts of odd specimens of the Human species'. Darwin's foray into the beer halls and gin palaces of East London might have been an unusual step for a gentlemen in the 19th century, but it enabled him to gather vital evidence of artificial selection in action. Natural science owes an important debt to amateur associations like these pigeon fancier clubs who continue to offer an important resource of expertise as well as a platform for the exchange of new ideas and the advance of scientific knowledge.

Portrait of a pouter pigeon, coloured engraving by D. Wolstenhome, made in 1862, from Darwin's home, Down House, Kent.

interspersed with hours spent prostrate on the sofa, plagued by bouts of violent sickness. There have been many theories over the years about the cause of Darwin's illness, including the suggestion that it might have stemmed from an inherited family weakness and the possibility that it may have been induced by a parasitic organism contracted during his travels around the world. One plausible explanation is that it was a manifestation of the terrible guilt and anxiety that he struggled with throughout his life. This was no dispassionate man, seeking facts to test a scientific theory. This was a man lured by ideas that contradicted the orthodox beliefs of the society that supported him and, even worse, the people he loved most — his wife Emma and dear friends John Henslow and Charles Lyell.

SHARING IDEAS

At the same time as wrestling with his conscience, Darwin delighted in the scientific importance of his work. He knew that in one sweep his theory took the accumulated facts of geology, palaeontology, comparative anatomy and the new life sciences, and explained them as a unified whole. One of the scientist's greatest motivations is the recognition of his or her work by the scientific community and, although he kept out of the public eye, Darwin couldn't resist sharing his ideas with a few carefully selected colleagues. He needed to test his ideas, to convince other scientists of the validity of evolution by natural selection and, in his increasingly invalid and reclusive state, to use them as foot soldiers to bring back information

Botanist Sir Joseph Dalton Hooker was the first person to whom Darwin showed an outline of his evolutionary theory.

to him from the real world beyond the walls of Down House.

Although Darwin nursed a dream of his theory becoming accepted by the established scientific community, he was happy to content himself with minor conquests, friends who could be trusted to keep a confidence. They included Charles Lyell, the botanist Joseph Dalton Hooker (1817–1911), who became the Director of Kew Gardens from 1866, and the enigmatic biologist Thomas Henry Huxley (1825–95) — the driving force behind the professionalization of British science that would influence the fate of The Natural History Museum in the second half of the 19th century.

BREAKING COVER

Darwin would probably have been content to keep his thoughts private, but great changes were in the air. First published in 1844, based

Key events that culminated in the publication of *On the Origin of Species by Means of Natural Selection.*

1753 Carl Linnaeus publishes *Species Plantarum*, which marks the beginning botanical nomenclature

1758 Carl Linnaeus publishes *Systema Naturae*, marking the beginning of zoological nomenclature

1797 Reverend Thomas Malthus publishes his *Essay on The Principle of Population*

1809 Jean-Baptiste Lamarck publishes *Philosophie Zoologique*

1812 Georges Cuvier promotes his opinions through 'Discours sur les revolutions de la surface du globe'

1830 Charles Lyell promotes uniformitarianism through his book *Principles of Geology*

1831–36 Charles Darwin's voyage on HMS *Beagle*

1844 Robert Chambers publishes *Vestiges of the Natural History of Creation* anonymously

1857 Richard Owen announces support for a limited process of transmutation controlled by God

1858 Darwin's and Wallace's papers proposing their theory of evolution by natural selection are read to the Linnaean Society

1859 Darwin publishes *On the Origin of Species*

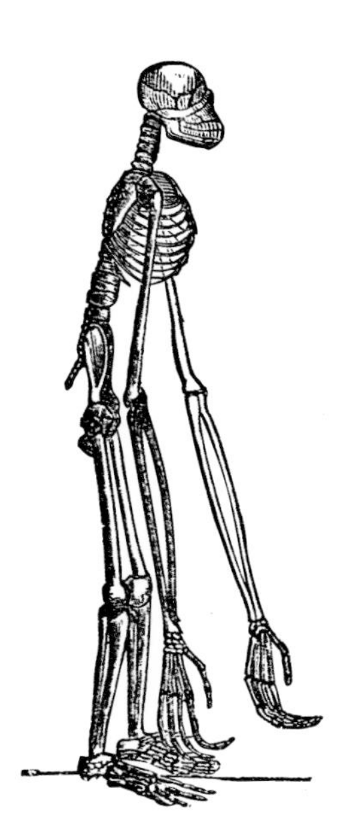
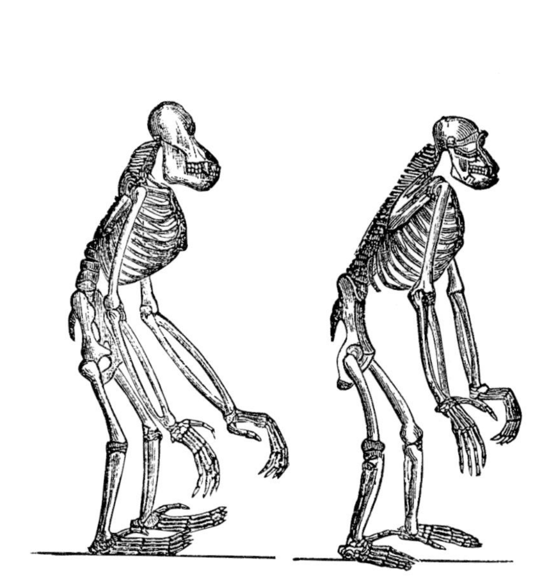
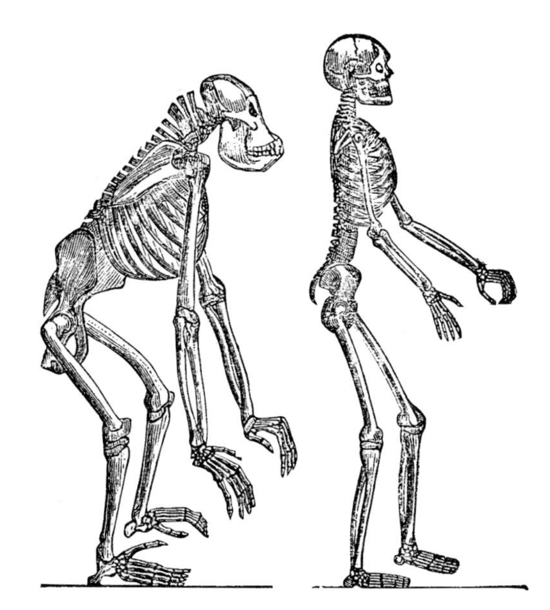

A provocative illustration depicting the monkey origins of humans, from T.H. Huxley's *Man's Place in Nature*, published in 1863.

loosely on scientific fact, an anonymous book, *Vestiges of the Natural History of Creation*, took transmutation out of the world of academia and popularized it for the benefit of the ordinary person. The book suggested that life had begun as an electrochemical reaction and, with guidance from God, simple organisms had undergone a process of progressive change that eventually led to the creation of humans.

The Church might have bristled at the link between animals and humans, and academics might have sneered at the book's scientific inaccuracies, but its message of divinely controlled progression and upward mobility made it an instant best seller. Although the initial hiatus surrounding the publication of *Vestiges* had encouraged Darwin to keep his ideas to himself, over time lengthy public debate created a more accepting climate that made him less nervous about going public.

Despite its success, *Vestiges* remained an anonymous publication until after the author, Robert Chambers, the Edinburgh writer, publisher and amateur geologist, died in 1884. Chambers was once asked why he never owned up to its publication. He replied 'I have 11 reasons', referring to his 11 children — an indication that although transmutation and evolution became more acceptable, they remained extremely sensitive issues in Victorian Britain.

By the 1850s, it was no longer taboo to discuss transmutation. Surprisingly, Richard Owen had as much to do with this changing attitude as Chambers' *Vestiges*. In 1849, the discovery of a new ape, the gorilla, in West Africa, had once again raised the question of the relationship between apes and humans. Rumours of its discovery became a frightening reality when the first animal was displayed to the British public, in 1855. Stories about the ape ancestry of humans started to circulate in the press, and the establishment turned to Owen to quash these morally subversive rumours once and for all. Owen, recognizing the turning tide of opinion, made a gesture towards transmutation and announced his acceptance of a 'limited' progression in nature. The progression was, of course, pre-ordained by God, and Owen remained steadfastly opposed to the idea of apes transforming into humans.

WALLACE'S CONTRIBUTION

In 1855, still having difficulty in accepting the idea of changing species himself, Charles Lyell spotted a paper entitled 'Introduction of Species' in the popular 'Annals and Magazine of Natural History'. The article had been written by the explorer and zoogeographer Alfred Russel Wallace (1825–1913) during his travels in Borneo, and hinted at some of the same ideas that Darwin had suggested to him about natural selection. Wallace was an impoverished land surveyor, who inspired by *Vestiges*, Darwin's account of his voyage, and Von Humboldt's popular travelogues, had set out with his friend the naturalist-explorer Henry Walter Bates (1825–92), first on a voyage to the Amazon, and later to the Malay Archipelago, to explore the concept of species. Exchanging land surveying for trophy hunting, he collected butterflies, beetles and bird skins for a dealer in London to finance his nomadic existence.

It was during the monsoon season, while lying in a hut on the Indonesian island of Ternate, suffering from malaria, that Wallace, like Darwin, took ideas from Malthus and used them to devise his own theory of evolution. He wrote 'At the time in question I was suffering from a sharp attack of intermittent fever, and every day during the cold and succeeding hot fits had to lie down for several hours, during which time I had nothing to do but to think over any subjects that particularly interested me.'

One question in particular returned to haunt Wallace: what was it that determined which particular individuals of an animal species, including humans, lived and died in a population? Eventually, he wrote 'And the answer was clearly, that on the whole the best fitted live. From the effects of disease the most healthy escaped; from enemies, the strongest, the swiftest, or the most cunning; from famine, the best hunters or those with the best digestion; and so on. Then it suddenly flashed upon me that this self-acting process would necessarily improve the race, because in every generation the inferior would inevitably be killed off and the superior would remain — that is, the fittest would survive.'

Although Lyell was struggling with the concept of evolution, he strongly urged Darwin to start preparing his theory for publication, suggesting that if he didn't some other young naturalist, such as Wallace, would. Darwin read Wallace's paper but, misinterpreting its guarded language, was convinced the author was just

THE WALLACE LINE

In many respects, Alfred Wallace has become one of the forgotten heroes of modern natural history. Most people associate evolution and natural selection with Charles Darwin, and are probably unaware that Wallace proposed this model for evolution, too. He made many other important observations during his overseas expeditions, and was the first person to propose an imaginary boundary line (subsequently named the Wallace Line), running between Borneo and Sulawesi within the Malay Archipelago, that separated Australian animals from those found in Eurasia. As Wallace wrote 'In this archipelago there are two distinct faunas rigidly circumscribed which differ as much as do those of Africa and South America and more than those of Europe and North America; yet there is nothing on the map or on the face of the islands to mark their limits. The boundary line passes between islands closer together than others belonging to the same group. I believe the western part to be a separated portion of continental Asia while the eastern part is a fragmentary prolongation of a former west Pacific continent'. Geological evidence gained in the 20th century shows that Wallace was very nearly correct in his assumptions. In fact, the eastern part of the area Wallace studied has since been shown to belong to the Australian continental plate, and it is customary now to think of the Wallace Line more as a narrow zone where Eurasian and Australasian organisms have met and mixed.

another creationist. He didn't want to rush his theory out in an insubstantial paper but, encouraged by Lyell's concern, began drafting a weighty tome, more appropriate for the announcement of a seminal theory. Hooker acted as referee on the slow succession of draft chapters, while the ever-cautious Darwin wrote to Wallace partly to discuss his paper, but also to stake his claim on evolution and subtly warn the young explorer off.

A JOINT DISCOVERY

If Darwin misinterpreted the article published in 'Annals and Magazine of Natural History', Wallace equally misinterpreted Darwin's letter. Flattered by the attention of his great hero, he completely missed Darwin's hints of ownership, and set about developing his theory in full. There were subtle differences between the two theories, but the overlap was undeniable. Wallace's paper on evolution by natural selection arrived in the post at Down House on 18 June 1858. Pipped at the post, and with his youngest son dying of scarlet fever, Darwin was inconsolable. He turned to Lyell and Hooker to ask for their guidance. Scrupulously honest, he volunteered to send the paper to a journal on behalf of Wallace for immediate publication but, with their prior knowledge of Darwin's work, both Lyell and Hooker agreed that the fairest solution would be a joint announcement of their discovery. Wallace was flattered, Darwin relieved, and on 1 July 1858 both papers were read by the secretary of the Linnaean Society to a meeting of its members.

Neither Wallace, on his travels abroad, nor Darwin was able to attend the meeting. By now, Darwin's illnesses and anxieties had become such a dominant aspect of his life that the time he was able spend away from Down House had become increasingly rare. After his years of tortured anxiety, the immediate reaction to the theory of natural selection was something of an anticlimax. There were no abusive comments or shouts of outrage. In fact, after the meeting, the president of the Linnaean Society remembered thinking that 1858 had not 'been marked by any of those striking discoveries which at once revolutionise, so to speak, [our] department of science'!

Alfred Russel Wallace, the collector who independently reached the same theory as Darwin while travelling in the Indo-Australian archipelago.

BEST SELLER

It was the publication of Darwin's book, *On the Origin of Species by Means of Natural Selection*, in 1859, that brought the theory into the open and triggered the real debate. Snapped up by the general public and scientific community alike, the book was an instant success, equalled only by sales of Lyell's *Principles of Geology*. The reaction was predictable, the radicals cheered, the orthodox community complained to Owen as president of the British Association for the Advancement of Science, and the growing body of middle-class naturalists, such as Thomas Huxley, welcomed Darwin's theory as a vital weapon in their campaign for the professionalization of science.

Huxley and Hooker were part of a new breed, disenchanted with the lack of professional opportunities for scientists in England. It was time for science to lose its amateur status and its outdated country-parson and gentleman-hobbyist image. It was time for an end to the established system of privileges that gave the Church its monopoly on science, and for the beginning of a new dynamic organization that would give scientists the opportunity to compete for research posts and teaching positions on merit.

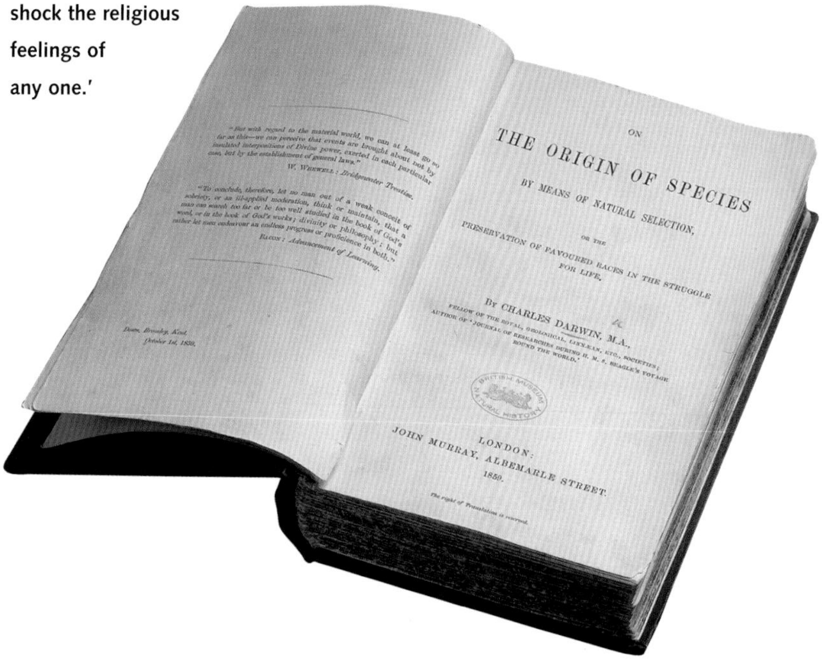

In his great book *On the Origin of Species* Charles Darwin wrote 'I see no good reason why the views given in this volume should shock the religious feelings of any one.'

A CAMPAIGN FOR SCIENCE

In 1856, Huxley, Hooker and others initiated a campaign to emphasize the practical benefits of science to society, and draw public attention towards the need for government to support and fund a national research programme. Hooker's programme at Kew Gardens for transplanting viable plants from their country of origin, and cultivating them in different countries within the British Empire, provided a perfect illustration of the economic benefits to be had from science. The cinchona tree, for example, grown to produce the anti-malarial drug quinine from its bark extract, became one of the programme's greatest successes when it was taken from Peru, its country of origin, and cultivated successfully in India.

This new utilitarian view of natural science provided a persuasive argument to justify increased government investment in science. However, professionals like Huxley still needed more ammunition if they were to undermine popular belief in divine creation, and overthrow the controlling influence of the Church once and for all. Evolutionary theories had been around for many years to challenge the concept of creation, but before Darwin's and Wallace's breakthrough, scientists lacked a suitable mechanism to explain how this process took place in the natural world. Natural selection seemed to offer a plausible explanation and, although Huxley at first doubted its truth, he seized upon the mechanism described by *Origins* to discredit the arguments put forward by orthodox naturalists such as Owen, who continued to deny the evolution of species and any link between humans and apes. Darwin's and Wallace's theory provided the legitimizing philosophy that Huxley needed to gain public support at all levels of society for the professionalization of science. The concept of evolution encouraged the less privileged members of society to seek self-improvement, change and progress, while natural selection, with ideas like 'the survival of the fittest', appealed to the industrialists and the nouveau riche, because it seemed to justify unregulated capitalism and the imperialist society it had generated.

The Plan is in the Detail 1860–1883

The debate about evolution and skirmishes with Huxley continued while Owen forged ahead with his grand plans for the new museum. In 1863, the House of Commons approved the purchase of 12 acres of land at South Kensington to accommodate the new building. Land was at a premium in Bloomsbury and, despite concerns about separating staff from one of their essential resources, the library at the British Museum, the decision was made to move the natural history collections to what was then a cheaper part of London.

In 1864, using Owen's original plans, the architect and engineer Francis Fowke (1823–1865) won a competition for the design of The Natural History Museum. Fowke died before work began, and in 1866 the Commissioner of Works awarded the contract to Alfred Waterhouse (1830–1905) instead. Although only 36 years old, the young architect had already gained a glowing reputation for his design of the Manchester Assize Courts and, while he was delighted to receive the museum commission, he was keen to inject his own style into Fowke's designs.

Alfred Waterhouse painted this watercolour of the final design of the Museum in 1876. The side façades were never completed.

Waterhouse was a Gothic revivalist. He was part of a new generation of architects and artists who despised the Renaissance style favoured by Fowke in his original plans. They believed the Renaissance represented a period of pagan immorality, in contrast to the far superior, deeply spiritual Mediaeval age. Waterhouse agreed with Owen that a national museum

WATERHOUSE'S ARCHITECTURE AND DECORATION OF THE NATURAL HISTORY MUSEUM

It was partly Richard Owen's desire to see the building decorated with animals and plants to symbolize the Museum's contents that enabled Alfred Waterhouse to argue a convincing case for changing the architectural style of the building. Elaborate carving and decoration was a feature of Romanesque rather than classical Renaissance architecture. The Natural History Museum was possibly the first building in the world to be faced entirely with terracotta. This material was a favourite with Waterhouse because it was cheap, durable and ideal for creating elaborate mouldings that exactly replicated the original design of the artist who sculpted them. It was also easy to clean, an important consideration for buildings in grimy Victorian cities like London. Using expert information supplied by Owen and his staff, Waterhouse made all the original sketches for the models of decorative plants and animals that were used to cast the figures in terracotta. These exquisitely sculpted figures were divided into living species, which decorated the zoological wing of the Museum, and extinct species, used to adorn the geological wing.

To this day, they continue to serve an educational purpose, as well as adding to the pleasure of walking a round the Museum. Although most of the terracotta sculptures represent examples from the animal kingdom, the ceiling of the Main Hall is devoted to plants. Consisting of 18 panels divided into nine lesser panels, 162 panels in total, it was painted by Messrs Best and Lea of John Dalton Street, Manchester, under the direction of William Carruthers, the keeper of Botany at that time. The stunning illustrations are heavily stylized, resembling artwork found in medieval manuscripts.

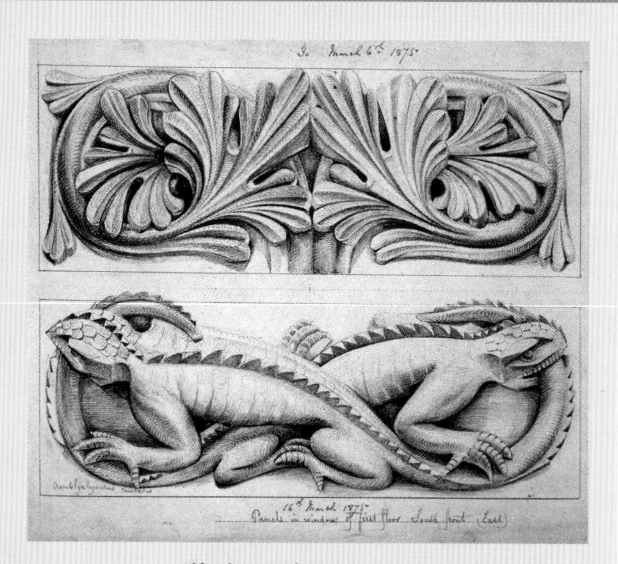

Alfred Waterhouse's pencil sketches of a monkey and a lizard for decorative terracotta panels in the Museum.

In the 18th century, physician and botanist William Withering investigated the therapeutic effect of foxglove, *Digitalis purpurea,* on patients suffering from dropsy (oedema), a symptom that can suggest a failing heart. This life-saving plant is one of the many plants that decorate the ceiling of the Waterhouse Building.

The plants depicted in the uppermost, unlabelled panels are difficult to identify, but the illustrations appearing two rows below become more accurate, and include the species name. The plants are mostly of medicinal or economic importance, including apple, aloe, tea, coffee, tobacco, Scots pine, strychnine, olive, cocoa and maize.

devoted to the divine glory of nature should in itself be a tribute to God, and he set about amending Fowke's design, creating a Romanesque style deliberately reminiscent of cathedrals that he had seen while travelling through Germany. He chose Romanesque because it was the style that had inspired Gothic architecture, but at the same time it was much closer to the classical architecture favoured by Fowke, and therefore required fewer modifications to the original plans that had already been approved for the building.

ABOVE: **Alfred Waterhouse's drawing for a decorative panel showing the dodo.**

OPEN AT LAST

Problems with funding, and difficulties with the supply of vast quantities of terracotta, needed to create the façade for the building, caused further delays until finally, 15 years later, in April 1881, The Natural History Museum finally opened its doors to the public. The new building met with a mixed reaction. The *Times* was full of praise: 'A true Temple of Nature showing as it should, the Beauty of Holiness' and 'The walls and ceilings are decorated as befits a Palace of Nature, with all the varieties of animal and vegetable life'. The experts were more critical. An article published in the august science journal *Nature* in 1882 argued that the style of the building was a mistake. It pointed out that the elaborate internal columns, arches and decorations interfered with the lighting of the specimens on display, and seemed to compete with the display cabinets rather than complement them.

The keepers weren't particularly happy with the new building, either, complaining that the recesses were badly lit and 'ill-adapted for exhibition purposes'. Neither were they pleased with Owen's plans for his 'Index Museum' in the main exhibition hall. The world had moved on since Owen's original plans for the Museum in 1858. Gone were the days when a museum attempted to display all of its specimens to the public. Thanks to John Edward Gray and the example set by William Henry Flower, conservator of the Hunterian Museum, the museum staff had already adopted the modern practice of separating key specimens for display from the bulk of the collections used for study. Owen's plans were outdated. Too old to fight this last battle, he contented himself with overseeing the transfer of the remaining parts of the collection from the British Museum. He eventually retired in 1883, at the grand old age of 80, received a knighthood for his services to British science, and enjoyed nine years of retirement before his death in 1892.

RIGHT: **Sir Richard Owen, portrait from the magazine *Vanity Fair*, 1 March 1873.**

Adapting to Evolution 1884–1900

On 14 March 1884 William Henry Flower, a confirmed evolutionist and close friend of Thomas Huxley, replaced Richard Owen and became the first director of The Natural History Museum. After completing a degree in medicine at University College London, he volunteered to serve as a surgeon with the British Army fighting in the Crimea. Suffering bitter deprivation with thousands of soldiers, many of whom perished from the freezing weather and starvation, Flower was eventually relieved of his duties on the grounds of ill health. He returned to London and worked at the Middlesex Hospital before he took the position of conservator of the Hunterian Museum in 1870.

Sir William Henry Flower.

Flower's tact in dealing with his staff, and the painstaking commitment he showed in every aspect of his work, made him a welcome replacement for Owen. A passionate supporter of Darwin, with a flair for creating informative and aesthetically pleasing displays, Flower took charge of the main hall and created a unique exhibition to illustrate natural selection and evolution. None of his keepers shared his belief in the theory of evolution, yet the display went ahead without a hitch, a lasting testimony to Flower's popularity and extraordinary powers of diplomacy. The main exhibition hall had finally been used to illustrate diversity but with a very different purpose from the one that Owen had intended. Flower's exhibition, ahead of its time, was a symbol of the dominant role that Darwin's theory would have in shaping natural science in the 20th century.

A PLACE OF WORSHIP

It took another 40 years of research and an understanding of genetics before biologists generally accepted that natural selection offered the best explanation of evolution, yet by the time of his death, in 1882, Darwin had become so popular he was buried in Westminster Abbey. Why was Darwin heralded as genius in the 19th century if at that time only a small fraction of the scientific community were willing to support his ideas of natural selection? One of the major factors was Thomas Huxley's formidable skill as Darwin's 'campaign manager'. Popularly known as 'Darwin's Bulldog', Huxley continually engaged in widely publicized debates with the great orthodox authorities of his day, promoting the strengths of Darwin's theory and keeping it in the public eye.

Whatever the reason for the delay, acceptance was only a question of time. Like Newton's laws in physics, the theory of evolution by natural selection drew together previously unexplained phenomena in taxonomy, geology and palaeontology, and explained them as an interrelated and unified whole. Naturalists were no longer restricted to descriptive activities. Darwin and Wallace had given them a wide-ranging theory that opened up a whole new world of testing and investigation. It was the starting point of investigations into heredity, and the point at which natural history ended and the modern natural and life sciences we recognize today began.

Like Newton's laws, evolution didn't necessarily mean an end to religion or a belief in God. In *On the Origin of Species*, Darwin was extremely careful to avoid the question of how life had originated. It was perfectly reasonable to continue to believe that God had created the Universe and life, but also to assume that once it had begun, the process by which species had changed and evolved were the result of entirely natural processes. The implicit challenge to the belief in divine creation was nevertheless a serious blow, and it marked the point at which the Church began to lose its monopoly on moral welfare and social control.

We began with The Natural History Museum building and it is a fitting place to conclude. The Museum was designed for an age when natural history was devoted to the glory of God yet, by the time it opened its doors, the revolution that would change our understanding to that of a world governed by natural processes was already underway. When the theory of evolution changed natural history to a science, the split with theology and the church was irrevocable. This explains why, in answer to the questions posed at the beginning of this chapter, many 20th century visitors are surprised that the Museum feels like a place of worship, but are unable to make an immediate connection between religion and the science on display today.

Natural Science

What is Science?

'Why sometimes I've believed as many as six impossible things before breakfast.'

Through the Looking Glass
Lewis Carroll

Understanding Science

How many of us really understand science? From *Frankenstein* to *Jurassic Park*, western culture is fed on a diet of novels, comics and films that perpetuate the image of naive or subversive scientists with secretive, often dangerous, motives. Is this disturbing vision anywhere near close to the truth, or is science more like the 'ivory tower' image that most of us were presented with at school — in which dispassionate, highly specialized experts, somehow isolated from the real world, test objective facts and theory with rigorous experiments? Looking into the private worlds of such eminent scientists as Charles Darwin and Richard Owen, we have already seen that, try as they might to be as objective as possible, their private beliefs play an influential role in how they interpret facts and construct theories. So how true are the other assumptions that we make about science?

Boris Karloff in Universal Pictures' 1935 version of *Frankenstein*.

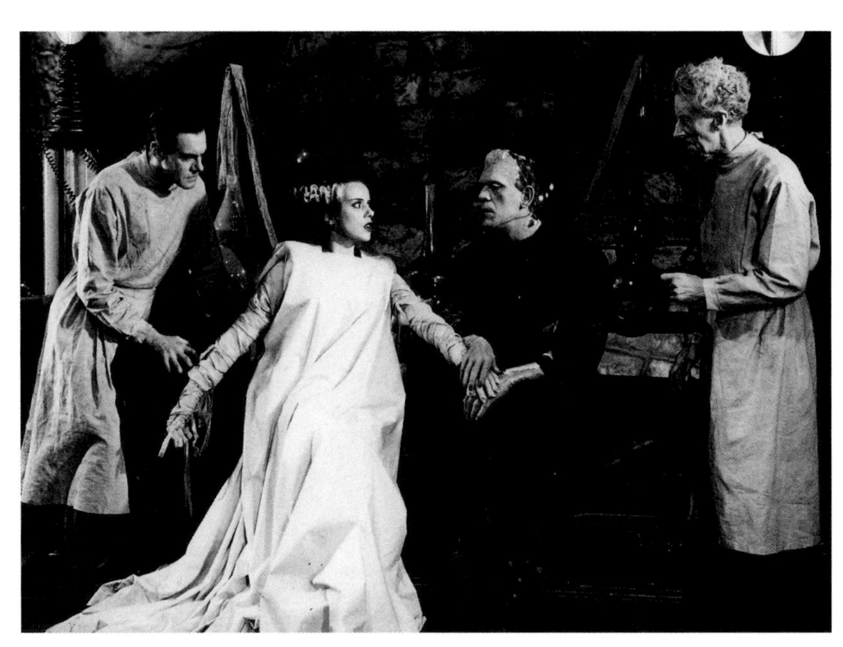

Using examples from both the past and the present day, this section takes common misconceptions and sheds some light on the real world of science. It won't turn you into a botanist or a palaeontologist, but it will help you reach a more balanced view about an activity that has an impact on nearly every aspect of modern life, through technological advances such as antibiotics and the computer, and through influencing the way we think about issues like the environment, nuclear warfare and human reproduction.

THE DIFFICULTY OF DEFINITION

Don't feel downhearted if you find it difficult to define science. Philosophers, historians and scientists have struggled for years to come up with a simple definition. So far, nobody has been completely successful. Even the brilliant Thomas Henry Huxley ran into difficulties when he described science as 'nothing but trained and organized common sense'. As Ernst Mayr, one of the leading zoologists of the 20th century, points out in his 1997 book *This is Biology: the Science of the Living World*, 'Common sense is frequently corrected by science. For instance common sense tells us that the Earth is flat and the Sun circles the sky'.

Although every branch of science is united in the common purpose of extending our understanding of the way the world around us works, it is difficult to make generalizations about disciplines that range from highly mathematical, theoretical sciences, such as particle physics, to observation-based sciences, such as taxonomy. A far simpler approach is to look at science as an activity, and find out what unites scientists in their exploration of the world.

THE FACTS NEVER SPEAK FOR THEMSELVES

Many people assume that science advances with the discovery of facts, but it is not so much the facts themselves as the interpretation of these facts — how they relate to one another — that leads to a progress in knowledge. Take the trivial example of the number of legs on a cow. If you've only ever seen one cow, the fact that it has four legs tells you only that the cow you are looking at has four legs. How would a scientist in the same situation interpret the evidence provided by this four-legged cow to increase his or her understanding of cows in general?

Science is about asking questions, identifying problems, making educated guesses to answer them, and setting about testing these guesses to see how accurate they are. In the case of the four-legged cow, the question might be 'How many legs do cows have', and a reasonable guess would be 'All cows have four legs'. These educated guesses are called 'hypotheses', and they lie at the heart of scientific enquiry. Hypotheses have been described as 'ideas on trial', because they enable scientists to test the accuracy of their hunches by using evidence from further observations or experiments. In this case, the obvious way for a scientist to test the hypothesis that all cows have four legs is to find more cows and count their legs.

Theories are educated guesses, too, but, in contrast to hypotheses, they are more comprehensive, supported by more evidence, and more widely accepted. Darwin, for

The sheeted breed of Somersetshire cattle.

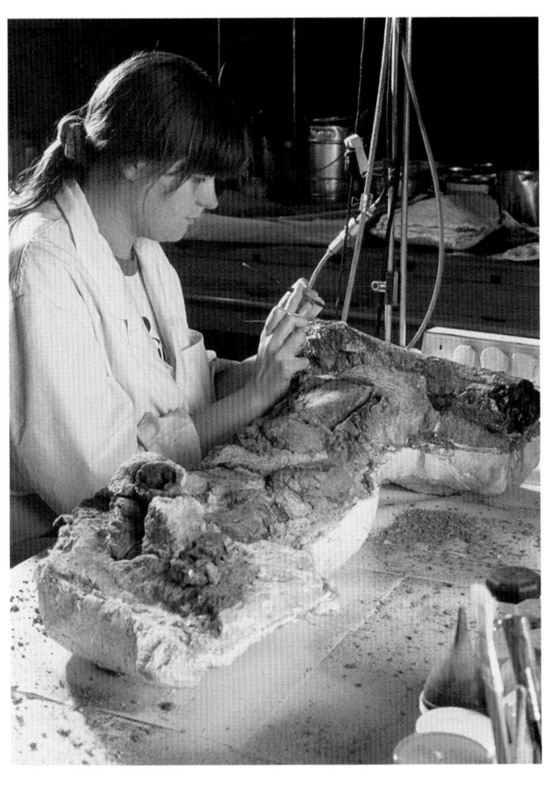

LEFT: **Scientist working in a palaeontology laboratory applying a resin solution to consolidate and protect freshly prepared fossil bones, here the ankle and foot bones of the dinosaur *Edmontosaurus*.**

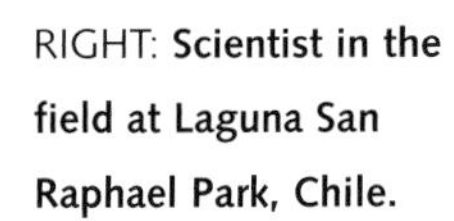

RIGHT: **Scientist in the field at Laguna San Raphael Park, Chile.**

example, developed and tested numerous hypotheses, accumulating a vast body of evidence to support his theory of evolution by natural selection. It took many years of scientists conducting independent research, and the discoveries of genetics, for his theory to gain the wide acceptance that it continues to enjoy today, albeit in a modified form.

TRUTH OR LIES

Scientists extend our understanding of the world by testing the explanatory power of hypotheses and theories. Because scientists are seeking the most accurate explanation of the world, many of us assume scientists accept scientific theories only if they believe them to be true. In reality, the opposite is the case. Scientists believe that all theories are flawed, but that some offer a closer approximation to the truth than others.

This approach might seem counter-intuitive but, if you stop to think, it makes perfect sense. If scientists believed theories were true, there would be no motivation to look for alternative explanations. Believing they could be false opens up a never-ending possibility for further tests and investigation. Inviting fellow naturalists to spend weekends at Down House, and picking their brains to uncover every possible argument against the concept of evolution, became a key tactic in Darwin's overall strategy. Exposing criticisms of his model — the lack of a suitable mechanism to explain the process of evolution for example — helped him to isolate weaknesses in his argument, and search for new evidence to overcome every possible line of attack.

Maintaining an open mind can be the greatest challenge for scientists, especially if they have invested years of their life developing ideas and accumulating evidence that appears to support a favoured theory. Richard Owen had great difficulty in accommodating the concept of evolution, while Darwin's notebooks, by contrast, provide a fascinating record of the struggle he had to maintain his objectivity and refrain from assuming that his theory provided an accurate explanation of variation before he had the evidence to support it.

How do scientists test their ideas?

'Scientists are people of very dissimilar temperaments doing different things in very different ways. Among scientists are collectors, classifiers and compulsive tidiers-up; many are detectives by temperament and many are explorers; some are artists and others artisans. There are poet-scientists and philosopher-scientists and even a few mystics.'

'There is poetry in science but also a lot of book-keeping.'

Sir Peter Medawar, *Plato's Republic*, 1982

Most of us assume that scientists use experiments to test their hypotheses and theories, but not all disciplines lend themselves to this type of investigation. Conducting experiments is a particularly useful approach for quantitative sciences, such as physics, where predictable phenomena like the movement of objects can be tested and measured over and over again, but they are less useful for historical sciences, such as geology. Reconstructing complex living processes or historical events, like the changing shape of the Earth's surface, is far more difficult. Also, because it is impossible to rule out factors such as chance, the results tend to be less conclusive. Who, for example, could predict the future of evolution with any certainty when random events — for instance, an asteroid impact — could trigger mass extinctions like the one that many believe ended the reign of the dinosaurs at the end of the Cretaceous?

As Sir Peter Medawar points out in the quotation opening this section, scientists use many different methods to investigate the world. At natural history museums where the goal is to record the diversity of life and explore the mechanisms that produce and maintain it, the experts' work tends to be investigative rather than experimental. The methods they use include taking samples and making detailed observations of nature, followed by careful analysis to interpret and understand their findings. Their scientific investigations might consist of chemical, numeric or taxonomic (classification-based) analysis, or a combination of several methods, to describe the materials from which the Earth and the rest of the Solar System are made, and to establish how different species compare and relate to one another.

EXPERIMENTATION

Although experimentation is rare, when 'historical' scientists encounter suitable problems, they can create models that

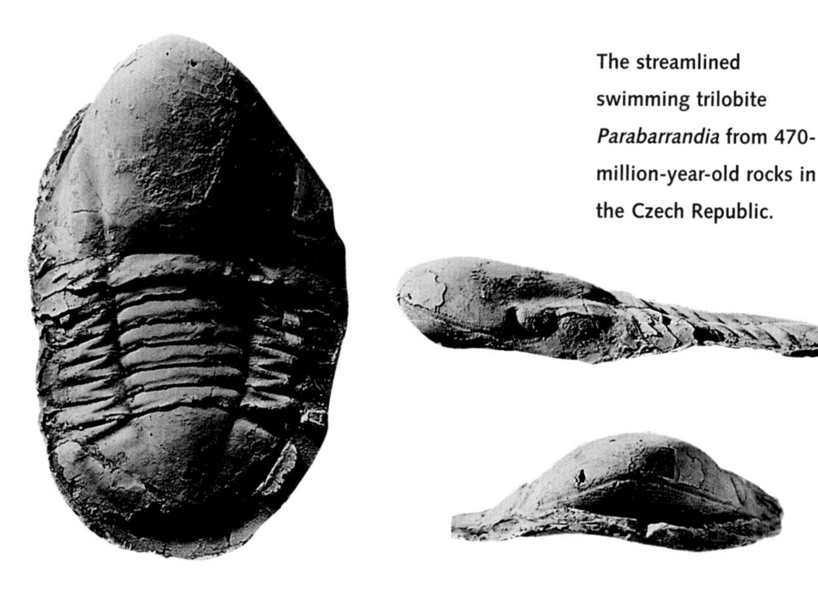

The streamlined swimming trilobite *Parabarrandia* from 470-million-year-old rocks in the Czech Republic.

reconstruct the past with great success. One of the challenges facing palaeontologists, for example, is the difficulty of forming testable hypotheses about the appearance, behaviour and lifestyle of organisms that have been extinct for hundreds of millions of years. Richard Fortey, a leading palaeontologist, shows how one ingenious experiment has helped us to understand more about trilobites — extinct, jointed-limbed invertebrate creatures that lived in the oceans from about 520 million to 250 million years ago.

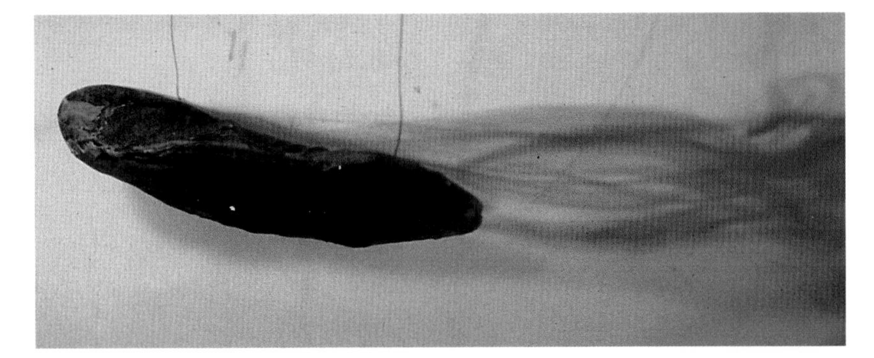

Flow patterns show the streamlined design of a model *Parabarrandia* swimming.

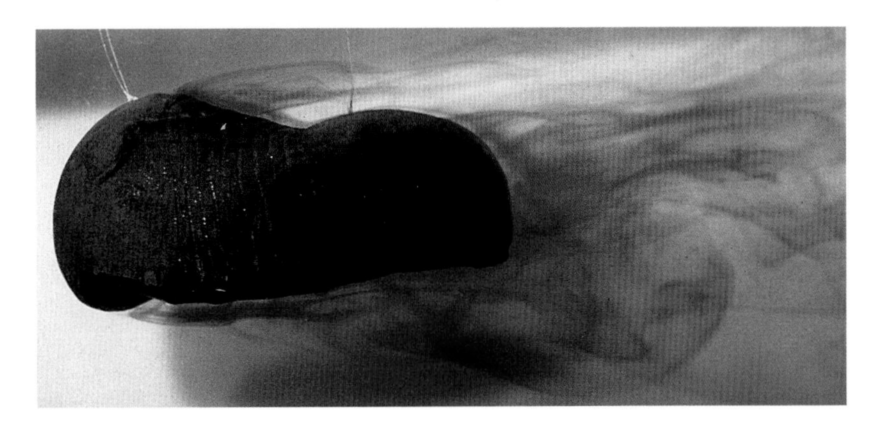

A poorly streamlined trilobite model of *Bumastes*, generates frictional drag and large wakes.

'Sometimes you can be working on a problem for years and no new insights emerge; on other occasions a bright idea can apparently spring spontaneously into your mind as if from nowhere. So it was with the case of the swimming trilobites.

I had been examining a fine cast of a large Czech trilobite called *Parabarrandia*, when I was suddenly struck by its resemblance to a dogfish. It seemed to have a long 'nose' at the front, produced by a projection of its convex head; at the same time its large eyes were tucked away so that they did not project like those of other trilobites. I rushed up to Imperial College to show it to a hydraulic engineer — he recognized the shape of this animal at once as a hydrofoil, a streamlined design. Could it be that we had here the first example of a trilobite designed to swim speedily through the open ocean?

Between us we designed an experiment to find out if the bright idea stood up to critical examination. Imperial College is full of equipment that can be used to make precise measurements of swimming objects. Our experiment was actually rather simple: we suspended various models of trilobites in a flowing current, and discovered the frictional drag for each one. We had to measure precisely the deflection of the strings on which the models were suspended. Most of our trilobites were very poorly streamlined — they were most unlikely to have been active swimmers. But the hydrofoil trilobite behaved very well under these conditions. It really did cut through the water as predicted. This meant that we now had to revise our ideas about the generally sluggish habits of these wholly extinct animals.'

EXPEDITIONS

Because much of the science carried out at natural history museums is based on observation and comparison, collecting new specimens in the field is a vital element of investigative natural science. Collecting requires technical expertise and a sense of adventure. Expeditions aren't always fraught with danger, but Geoff Boxshall illustrates how adventurous contemporary collecting can be on a recent expedition to Siberia:

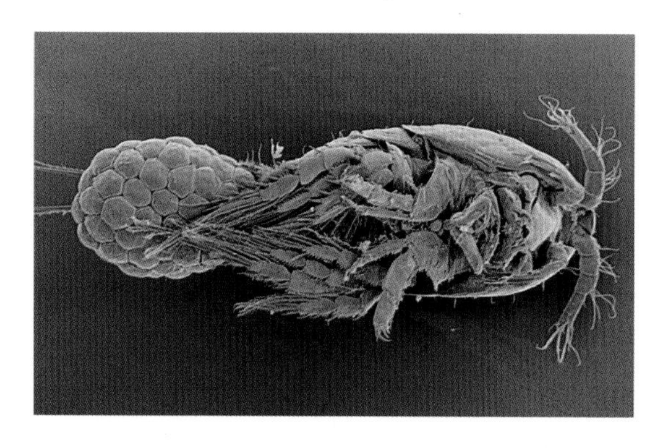

Copepod specimen (*Harpaeticella inopinata*) from Lake Baikal.

'While copepod hunting in Lake Baikal in Siberia, I was lucky enough to get a dive in a PISCES submersible. It was somewhat cramped with the pilot, the engineer and me all in the 6 ft diameter pressurized chamber inside the submersible, but I was too excited about the dive to notice. After the final technical checks, we began our descent, sinking rapidly until the last traces of the surface light vanished. After some more minutes falling into the blackness, the pilot switched on the external lights. I was amazed to see fishes, mostly the endemic Baikal oilfish, all standing on their heads. It took me a while to realize that they were following our lights down — a reminder that the mere presence of an observer can change the behaviour of wild animals. Our descent slowed as we reached the bottom. Out of my porthole I could see that the bottom, which I expected to be a grey desert, was alive with life and colour. There were purple and orange flatworms over 6 inches long, amphipod shrimps with immensely long spidery legs tiptoeing across the mud, and others shooting rapidly back into their burrows. Using the robotic arms of the submersible, we took samples with corers, nets and slurp guns. Thanks to modern technology, plus three summers' hard collecting around the shores of the lake, our museum now contains what is probably the world's most comprehensive collection of Baikalian copepods.'

COLLABORATION

Scientists have become increasingly specialized in the 20th century as a dazzling series of discoveries — ranging from Einstein's theory of relativity and the understanding of radioactivity to the discovery of the gene and the structure of DNA — has led them deeper and deeper into the complex mysteries of nature. So little was known of the world during the age of the Greeks that the great philosophers, such as Aristotle, could justly be regarded as masters of astronomy, mathematics and natural history. Today, the sheer volume of literature being produced within the individual fields of a discipline like biology makes it unlikely that experts could straddle another field within the same discipline, let alone another discipline altogether.

Increasing complexity has led to a trend towards teams of scientists working together, across disciplines and across nations, to solve the latest puzzles — rather than individual scientists, such as Einstein or Darwin, making great discoveries. New disciplines have continued to emerge over the last 100 years, and, in some instances, established disciplines have pooled resources to create hybrid disciplines. Cell biology and chemistry, for example, merged to create biochemistry.

Although some scientific work is still carried out by individual scientists working in solitude, particularly in the more theoretical disciplines,

Benedict Allen

MODERN-DAY EXPLORER

To start from first principles, to me exploration isn't about many of the sorts of things we have come to expect: conquering natural obstacles, planting flags or even science, if that only means extending our rationalized, Western understanding of the world. In other words, it's about leaving our world behind in order to encounter another. It's not about going where no one's gone before in order to leave your mark, but about the opposite of that — about making yourself vulnerable, opening yourself up to whatever's there and letting the place leave its marks on you.

In short, the word explorer has been misappropriated, or at least mislaid. It's become associated with the past, not the present, and with people who might be described as conquerors as much as discoverers. But whether we are Iban hunters of Borneo or Zurich bankers, we are all explorers. Our desire to see what's around the corner is there from birth.

Opening yourself up to another world in Machu Picchu, Peru.

'Explorers' on the trading floor of the Chicago Board of Trade.

Submersible being lowered for use in the study of the seafloor.

such as quantum physics, most modern research is conducted in teams. John Lambshead, a marine biodiversity specialist, explains why collaboration is an important aspect of deep-sea biodiversity studies:

'The deep sea is defined as the seabed beyond the continental shelf, i.e. in water deeper than 200 metres. This is roughly half the Earth's surface. Teamwork is essential for two reasons. Deep-sea research requires the use of large and expensive platforms. Just as space scientists need a rocket to carry out space research, marine scientists need expensive purpose-built research vessels for deep-sea science. These ships carry a wide range of equipment, including core-samplers, robot landers, remote-controlled vehicles and deep-diving submersibles. No one scientist can find sufficient funds to cover these costs, so groups of scientists club together to pool resources. Many of the questions now addressed can only be answered by multidisciplinary teams. For example, a project to analyse deep-sea biodiversity would include biodiversity ecologists, a number of taxonomists expert on the different groups of animals, DNA specialists, biochemists, chemists, oceanographers, and engineers. Deep-sea biodiversity scientists study the patterns of species diversity in the oceans, and try to understand the processes involved by relating animal patterns to environmental information about the physical world. The scientist tries to distinguish the signals left by historical evolutionary trends from modern ecological processes — what processes create diversity, and what processes destroy it.'

Pure and Applied Science

IS SCIENCE THE SAME AS TECHNOLOGY?

The simple answer is no. Science is an activity devoted to understanding and explaining how the world works. Technology refers to the practical application of this knowledge: the contraceptive pill, for example, is an application of scientific understanding of the human reproduction system. Advances in technology can also result from the development and modification of practical techniques. The English inventor Thomas Newcomen (1663–1729), for instance, invented the steam engine by trial and error, completely unaware of the scientific laws of thermodynamics that governed its action. Likewise, Hans Lippershey (*c.*1570 – *c.*1619), the Dutch spectacle manufacturer, to whom many attribute the invention of the telescope, knew little of the science of optics. Thanks to the telescope, Galileo and Kepler were able to improve our understanding of astronomy — showing that science not only encourages the development of technology, but benefits from it, too.

WHAT IS THE DIFFERENCE BETWEEN PURE AND APPLIED RESEARCH?

Research projects tend to be classified as either pure or applied science. Pure science is motivated by the quest to understand the world for its own sake, whereas applied science is motivated by the need to solve a practical problem. For example, some dipteran specialists study Diptera (flies) to understand them from an evolutionary standpoint in their own right, whereas others, such as malaria-focused dipteran specialists, study them to understand their role in the spread of malaria. Understanding the behaviour and.habits of the mosquitoes that spread malaria enables specialists to identify methods that will help to prevent and control fresh outbreaks of the disease.

In practice, John Lambshead points out, there is often considerable overlap between pure and applied science, and many scientists find themselves investigating both sets of problems.

'A pure-science researcher asks questions about the natural world, tries to find answers, and then publishes the results in scientific journals. Then, scientists working in other fields read the results and incorporate the data and conclusions into their own investigations. For example, a paper on the processes that control biodiversity might be used by an applied scientist developing pollution monitoring tests or a mathematician modelling the impact of global warming on the biosphere.'

PURE SCIENCE — CELL WALL DEVELOPMENT IN DIATOMS

Diatom expert Eileen Cox gives examples of the pure and applied research problems she is currently investigating.

'Diatoms are a cosmopolitan group of microscopic unicellular algae, which are distinguished by their highly structured, silicon-based cell walls. Although the precise cell wall structure has formed the basis of diatom classification for over a hundred years, how the walls are laid down has been studied in only a very few species. This means that diatom experts do not really know how the features they use to discriminate between species are produced. Cell wall development

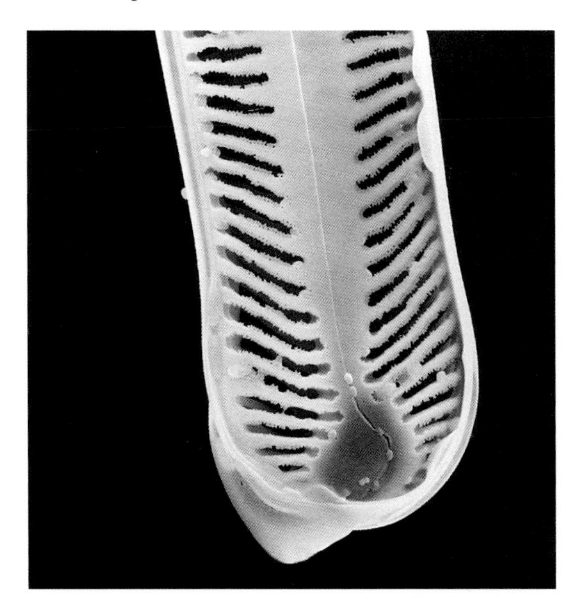

LEFT and RIGHT: **Cell wall diversity in diatoms.**

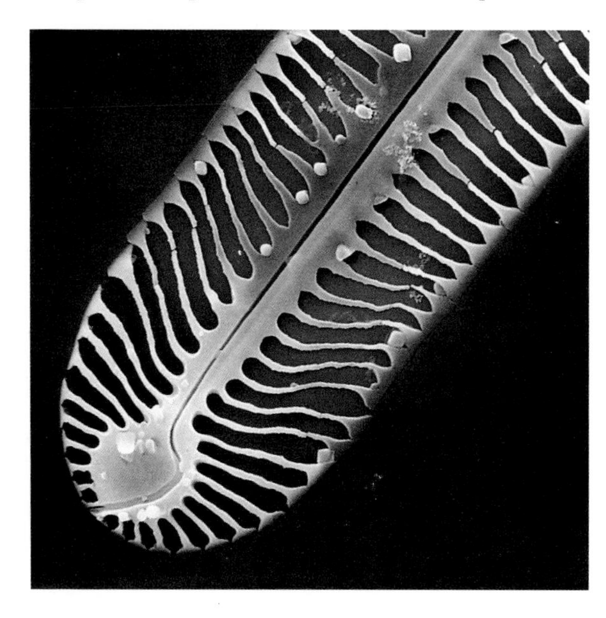

is now being investigated in a range of species to discover whether similar structures are always formed in the same way, or if different developmental pathways can produce similar end results.

Laguna San Rafael National Park, Chile.

Evolutionarily related species would be expected to share the same developmental pathways, whereas contrasting pathways producing similar structures would be considered the result of convergent evolution (in which unrelated species evolve similar adaptations as a result of living in similar environments) rather than relatedness. There are, therefore, two major aims: to determine precisely how different wall structures are formed, and to test the current hypothesis of relationships within the diatoms using this new information. In the future, because cell wall structure is under strong genetic control, it may be possible to use organisms such as diatoms to study how cells control the deposition of inorganic compounds to form complex structures. Small size, rapid growth rates, and the possibility of producing large numbers of genetically identical individuals favour the study of diatoms as model organisms for more general biological phenomena.

APPLIED SCIENCE — DIATOMS AND ENVIRONMENTAL CHANGE

Ecotourism and wide-ranging pollution are starting to have a negative impact on even the most remote areas of the world. Diatoms are sensitive indicators of environmental conditions, and have revealed the effects of acid rain, sea-level changes, climate changes, and eutrophication, in different parts of the world, but particularly in the northern hemisphere. In many cases, changes have already occurred, and baseline conditions must be reconstructed from fossil remains.

The opportunity to collect diatoms in the Laguna San Rafael National Park, in southern Chile, one of the most remote and unspoiled regions in the world, means that baseline conditions can be documented from the diatoms living there today, before the area is subjected to disturbance. Once the scientists have described the diatom populations of the clean lake and river systems, they can devise routine sampling programmes to allow local staff to monitor conditions at appropriate intervals. They can then assess any changes in diatom species by monitoring their ecological response, and implement management strategies where necessary.'

WHY FUND PURE RESEARCH?

The benefits of applied research are immediately obvious but, to many people, it is not clear that pure science serves any practical purpose, beyond extending our understanding of apparently obscure areas of nature. This short-term view fails to take account of the fact that most of the practical applications of applied science and technology are based on pure science discoveries that took place many years earlier. The discovery of X-rays in 1895 by German Nobel-Prize-winning physicist Wilhelm Röntgen (1845–1923), for example, led to the invention of medical X-ray technology soon afterwards. It also encouraged research into medical imaging that led to the development of computer-based radiography in the 1960s and three-dimensional scanning imaging, used today.

No-one can predict where pure science research will take us in the future, but the potential benefits are immense. Applied research can solve many immediate problems, but pure research holds the key to completely new horizons of opportunity.

WHY DO SCIENTISTS DO WHAT THEY DO?

Ask most scientists this question and they will say they do what they do because they enjoy it. Some scientists working for private companies might be motivated by financial reward, but most are drawn by a combination of other factors: the desire to solve a puzzle, a genuine concern about the world, the wish to make a difference and directly benefit others, professional prestige, or the lure of making an important discovery. The American molecular biologist James Watson who, along with the British Francis Crick, was attributed with the discovery of the structure of DNA, conveys the excitement of the race that preceded their ground-breaking achievement:

'Then DNA was still a mystery, up for grabs, and no-one was sure who would get it and whether he would deserve it if it proved as exciting as we semi-secretly believed.'

James D Watson, *The Double Helix*, 1968

More often than not, scientists are drawn by a particular problem that remains unsolved. To quote James Watson again, from the same source: 'My interest in DNA had grown out of a desire, first picked up whilst a senior in college, to learn what the gene was.'

In the case of the experts working at The Natural History Museum, motivation often stems from the desire to obtain as complete a picture as possible of one particular group of organisms or geographic area.

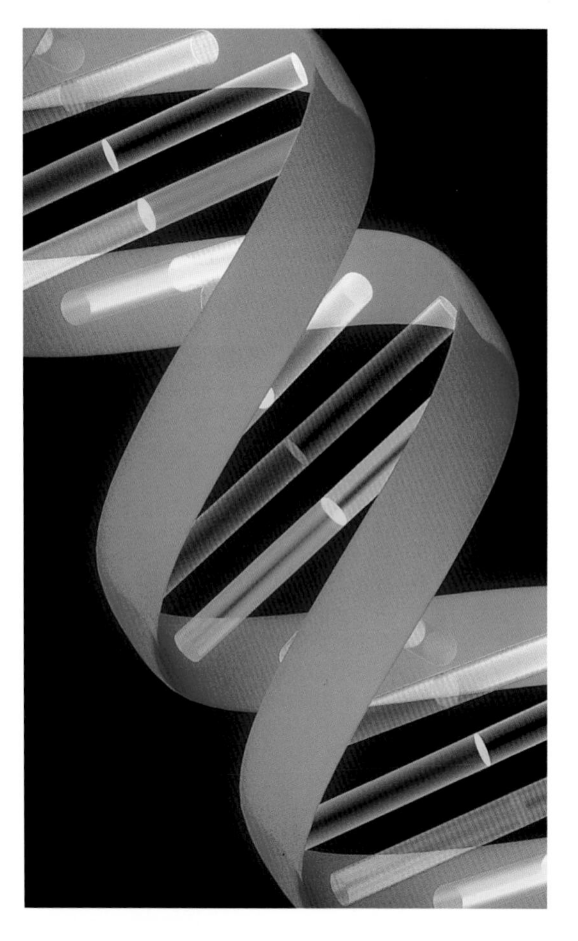

Model of the DNA double helix.

DISCOVERIES

'A discovery is like falling in love and reaching the top of a mountain after a hard climb all in one, an ecstasy induced not by drugs but by the revelation of a face of nature that no one has seen before.'

Max Perutz, Nobel-Prize-winning biochemist, quoted in Neil Campbell, *Biology (2nd edition)*, 1990

WHAT IS THE RELATIONSHIP BETWEEN SCIENCE AND RELIGION?

*'Science without religion is lame.
Religion without science is blind.'*
Albert Einstein

Dr. Robin Cocks, palaeontologist and President of the Geological Society of London explains why he believes that science and religion are not mutually exclusive:

Q What are your religious beliefs?
I am a Christian, and an active member of the Church of England.

Q As a scientist, do you find it hard to reconcile your religious beliefs and your science?
The simple answer is no, and there are two chief reasons for this. Firstly, science can never prove or disprove religion. Scientists require concrete evidence to prove or refute hypotheses and theories and no scientist has provided concrete evidence that God does not exist. Even putting my Christian faith to one side, as a scientist it seems sensible that I should keep an open mind on whether God might exist. Which is why I never cease to be surprised when I talk to scientists who categorically refuse to believe in the existence of God. For me there is no conflict. Science and religion can never prove or disprove each other.

Which brings me to the second reason, the concept of 'Occam's Razor', or the discipline of keeping things simple until they are proved more complicated. I believe that God, in his infinite wisdom, set up the physical world, created atoms and molecules and composed natural laws like gravity and evolution, and once established, left these underpinning processes to run themselves. Why would an omnipotent being need to be involved in the background running of the Earth and the Universe? Surely he prefers to influence less automatic events?

We don't need to believe in God to study and understand the wonderful complexities of the natural world, but we don't need to deny his existence either. To me it's a completely unnecessary risk. Unless there is final irrefutable proof, isn't it better to believe in God and live according to his guidelines, rather than face the possibility of his displeasure?

Q Has your research into the evolutionary history of seashells in the Ordovician and the Silurian periods ever challenged your religious beliefs?
I see no reason why the concept of evolution should challenge my religious beliefs. The record that we have for the evolution of seashells during the Ordovician and the Silurian periods, which together lasted over 75 million years and ended over 400 million years ago, provides a complete suite of specimens, with no missing links, that reveals an undeniable history of progression. For me, it provides irrefutable evidence of a wonderful, natural process that can be best attributed to the ingenuity of God. If anything, my research has strengthened, rather than questioned, my faith.

PASSIFLORA

Although some discoveries are more trivial than others, and not all scientists experience the ecstasy described by Max Perutz, it is obvious from listening to them and reading personal accounts that they derive enormous pleasure from making breakthroughs that contribute to our understanding of the world. Take the case of Sandy Knapp, a botanist, and her discovery of an unknown species of passion flower while collecting in Panama:

'The sheer rush of adrenalin as a botanist sees something she knows is new to science is indescribable; any comparison falls far short of the real emotion and excitement of the moment. While collecting in Panama, in a well-collected area mind you, my companion brought me a flower he had found floating in the river. To my delight I recognized it as an unknown species of *Passiflora* — and a fabulously beautiful one at that. The flower was large and creamy white, with a sweet smell.

Well, since it had come down the river, it just had to be upstream, so upstream I went, wading through the middle of the river, up to my waist in quite cold water.

Eventually, I found the plant, a large woody vine full of flowers in a large tree overhanging the bank. The tree was too high for me to climb (I'm not very good at climbing anyway!), so I found a man with an axe, who was living along the other bank, to come and cut off one of the branches so I could get at my prize. The branch fell with a crash into the river — whereupon I scrambled, getting wetter and wetter, to collect as much material as I could. If you cut down part of a tree it is pointless to leave any collectable material behind — it will not grow again and it seems a shame to go to all that effort for only a little gain. I hauled all my pieces of vine, tree and all the other epiphytes that were growing along the branches to the shore and began to press so that the flowers would not shrivel and make ugly specimens. I pressed and pressed, in a mad race against time.

Passiflora.

Sunset in the tropics is not the long-drawn-out affair of an English summer — it comes with a bang — and, much to one of my companion's annoyance, we were caught. Not only did we have to carry huge amounts of pressed material out for several miles, but we had to do it in the dark! To this day, I think he has never forgiven me. But I got the *Passiflora* — and described it as new, based on the material I so excitedly collected. The sense of satisfaction was immense, and I remain convinced that only the adrenaline of having found the new species got us back to the car through the dark.'

Of course, not all discoveries are the end result of tightly controlled investigations and chance; luck and flashes of inspiration can have as much to do with important break-throughs as any amount of careful planning and investigation. For instance, in 1929, Scottish bacteriologist Sir Alexander Fleming (1881–1955) discovered the antibiotic effects of the mould *Penicillium notatum* by accident as the fungus grew on unwashed apparatus in his laboratory. Australian pathologist Howard Florey (1898–1968) and German chemist Sir Ernst Chain (1906–79) successfully isolated penicillin, the active ingredient, ten years later, and America made it the world's first mass-produced antibiotic during World War Two. Fleming, Florey and Chain shared a Nobel Prize in 1945 for their achievement.

A NEW CLASS

The discovery of a new mammal or bird species is a remarkable event, but new species of aquatic invertebrate animals, such as crustaceans (crabs, shrimps and relatives) are relatively commonplace. The 42,000 species of crustaceans so far described by scientists probably represent only one-third of the real diversity of the group. Sometimes, very different species are discovered that have to be placed in a new family, or even in a new order (the next category up the scale). Rarely, however, is a new *class* revealed. This rarity isn't surprising since birds, mammals and reptiles are all classes of vertebrates, for instance, and it isn't often that such a new major category of animal gets discovered. In 1981, the crustacean world was stunned by the discovery of the remipedes. The Class Remipedia contains slender crustaceans, about 2-3 cm long, with rather centipede-like, segmented bodies bearing many pairs of swimming legs. They are all blind and white in colour, and each one of the 12 species currently known lives in flooded marine caves in coastal sites dotted erratically around the globe. They are not common and favour places where they are not easy to collect.

Remipede.

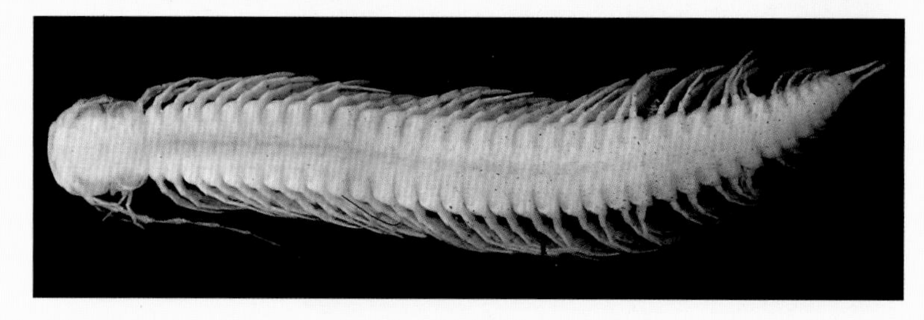

IN SEARCH OF REMIPEDES

Occasionally, the excitement surrounding a discovery can have even the most brilliant scientists jumping to the wrong conclusions. Geoff Boxshall describes the pitfalls of searching for remipedes (small, many-limbed crustaceans) in flooded caves on one of the Canary Islands:

'Jameos del Aqua is a lava tube on the island of Lanzarote, formed when the Coruna volcano erupted about 7000 years ago. It extends from the old volcano to the coast and out 1.5 km under the sea. The seaward end is flooded with seawater and is home to a special community of animals (the anchialine fauna), which exists only in a few similar, flooded coastal caves dotted around the world. This particular lava tube was of interest since it was the only place even near to Europe where remipedes had been found.

Jan Stock, a Dutch professor, Elias Sanchez, a Spanish student, and I were hunting for remipedes and other anchialine crustaceans around the Canary Islands. We had been searching caves and lava tubes on Tenerife and El Hierro, and had flown over to Lanzarote especially to visit Jameos del Aqua. We'd had a good day collecting the blind white crabs and small copepods (but no remipedes), when a chance conversation with a waiter in a local bar sent us to a nearby lava tube, Jameos de los Lagos.

We arrived at the entrance, a vertical hole about 8m across and 8m deep, and we could see the blackness of the cave leading off towards the coast. I was elected to go down to reconnoitre before our full-scale expedition the next day. Jan and Elias gave me 30 minutes to go down and check for water at the end of the cave.

I went steadily downwards, scrambling over boulders wherever part of the roof had collapsed. The joy of lava tunnels is that they go straight, and you can't easily get lost like you can in the labyrinthine tunnels of anchialine caves cut into coastal limestone rocks. After twenty minutes, there was still no sign of water and, as I would already be late back to the surface, I decided to turn back — but only after one last look over the mass of boulders immediately in front of me. Climbing to the top of the boulders, I could see my helmet light reflecting on water. There really were lakes (*lagos* in Spanish) in the Jameos de los Lagos!

I clambered down to the water's edge to taste whether it was seawater. It was. I turned off my light, and sat in the silent blackness for almost half a minute, beginning to focus my mind on the return journey. When I twisted the light back on, I saw, trapped in the pool of light just below the water's surface, a white, swimming, centipede-like animal about 2cm long — my first remipede? I grabbed a bottle from my rucksack and, with some luck, I managed to coax the animal to swim into the bottle. Screwing the lid back on, I sat down and counted to ten. I had to get back to the surface as my thirty minutes were up, but I had to go carefully with my precious find.

It was dusk when I finally emerged. Jan's complaints at my overdue return immediately turned to eager expectation when I held up the bottle, saying I thought I'd caught a remipede. Jan, the best field biologist I have ever encountered, produced a small dish and a hand lens from his voluminous pockets, and began to examine my prize catch. He turned with a smile, and said "I don't know what the British Museum [The Natural History Museum] is coming to these days when its staff can't even tell a worm from a crustacean". I had caught a beautiful specimen of *Gesiella jameensis*, a specialized polychaete worm found only in anchialine caves, but definitely *not* a remipede. Jan rarely let me forget!'

***Gesiella jameensis*, a polychaete worm, from Jameos de los Lagos, Lanzarote.**

Cave lake inside Jameos de los Lagos, Lanzarote.

NEAR MISSES

As we have already seen in the case of Darwin, Wallace and the theory of evolution by natural selection, it is not unusual for different scientists to make the same discovery independently of one another. Sandy Knapp explains how she was pipped at the post in naming a new species of epiphytic shrub (one that grows on another plant but is not parasitic on it) from Ecuador, but was fortunately able to turn this near miss to her advantage:

MARKEA SPRUCEANA

'Finding and describing a new species is always exciting — but one worry a taxonomist always has is that someone else has got there first, or will be doing it at the same time. This is usually not a problem, as there are so few taxonomists and so much diversity, but the possibility is always there.

I had described several new species of the genus *Markea* in the Solanaceae (the potato family), a group of spectacular epiphytic shrubs with big showy flowers, as part of the work associated with a checklist of the flora of Ecuador. Richard Spruce, one of my botanical heroes, in the mid-1800s had collected one of these — so I decided to name it in honour of him. Imagine my surprise, as I was correcting the page proofs of the article (the final step before publication), when I saw published in another scientific journal a plant with the name *Markea spruceana*! It was based on the same material as I had used, and so my description was superfluous.

Fortunately it was not too late for me to change my paper to amplify the description of *Markea spruceana* (I had seen better collections and could describe the fruit as well as the flowers), rather than describe it as new. It was a near miss from publishing something that would have sunk into obscurity instantly!'

Markea spruceana.

EVER-CHANGING SCIENCE

Our understanding of the world today is only as accurate as the available evidence allows it to be, and new discoveries lead scientists to constantly change and modify their views. In South Africa, for example, the discovery of unusually complete hominid remains, thought to be about three and a half million years old, will hopefully provide scientists with vital new information regarding the evolution of humans. And in the following example from botany, Sandy Knapp describes the recent discovery of a new family of plants in South America that completely overturns scientists' assumptions about the organization of sexual characteristics in flowering plants.

DISCOVERING A NEW FAMILY

'Discovering new species is exciting, but imagine discovering a plant that turns out to be an entirely new family! Esteban Martínez, of the Universidad Nacional Autónoma de Mexico, discovered just such a plant while collecting in the southern Mexican state of Chiapas. He collected a tiny thread-like saprophyte — a plant that does not produce its own food, but lives on dead and decaying organic matter — with some unusual features. Rather than having the ring of male stamens outside the female parts of the flowers, this tiny plant had the female parts surrounding a central group of stamens! This is known in no other flowering plants, and for this reason this tiny, apparently insignificant plant was accorded the status of an entirely new family. Martinez, and his colleague Clara Hilda Ramos, named it *Lacandonia schismatica*, the sole member of the family Lacandoniaceae: *Lacondonia* in honour of the Lacandón region of Chiapas where it was found, and *schismatica* (from the Greek *schisma*, 'a split') for the problems its unusual flowers caused for systematics. It was known from a single threatened population, but recently has been found in several other localities — however, its conservation status is by no means secure.'

This scanning electron micrograph of *Lacandonia schismatica* shows a longitudinal section through a flower, with two central stamens (male parts) surrounded by numerous separate pistils (female parts).

Reconstruction of hominid foot based on fossilised remains discovered in South Africa 1999.

STEADY PROGRESS AND REVOLUTION

Although most of us think of science advancing rapidly, entirely as the result of groundbreaking discoveries made by individual scientists like Newton and Darwin, in reality the process is more gradual and elaborate. Thomas Kuhn (1922–1996) a leading American philosopher and historian of science, described a cycle consisting of periods of steady progress, using 'normal' problem-solving, in which researchers test and refine a widely accepted theory, followed by periods of turmoil, or 'revolutions',

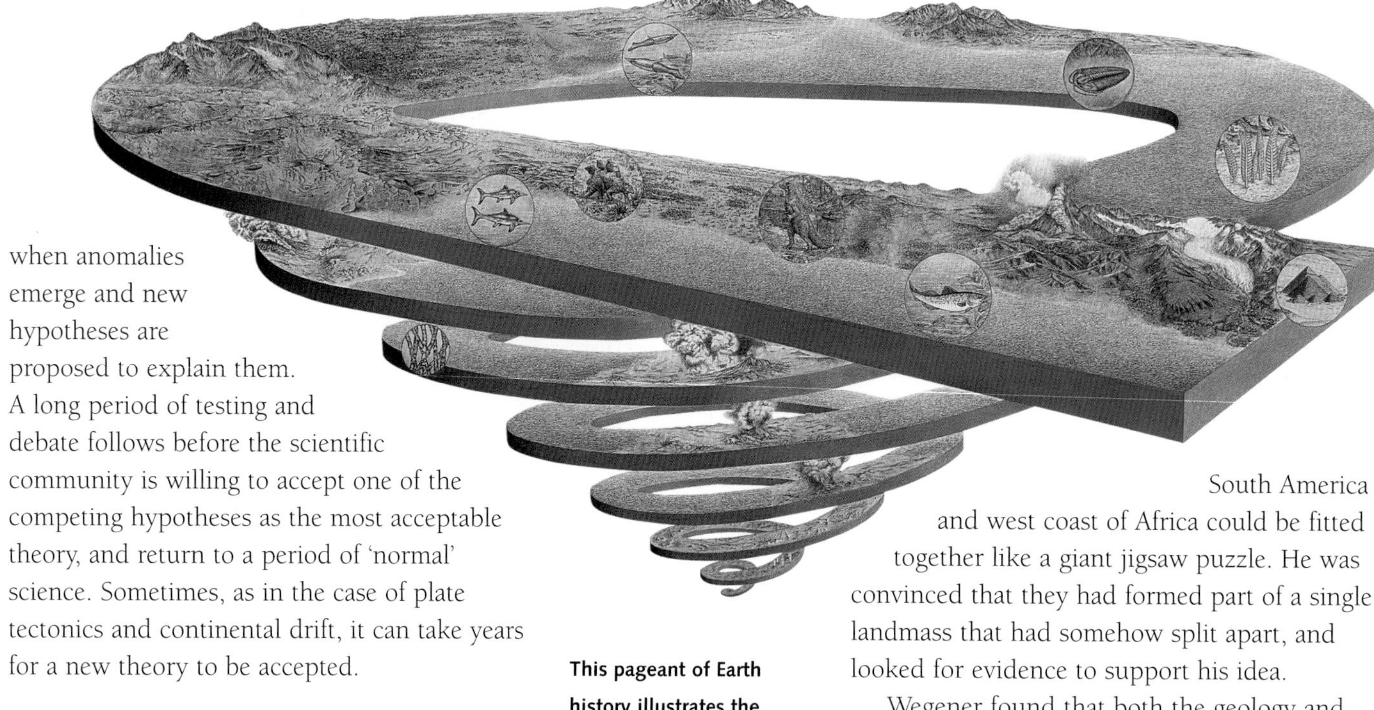

when anomalies emerge and new hypotheses are proposed to explain them. A long period of testing and debate follows before the scientific community is willing to accept one of the competing hypotheses as the most acceptable theory, and return to a period of 'normal' science. Sometimes, as in the case of plate tectonics and continental drift, it can take years for a new theory to be accepted.

This pageant of Earth history illustrates the sweep of geological time as a pathway and highlights some of the crucial events in the evolution of the Earth.

PLATE TECTONICS AND CONTINENTAL DRIFT

By the beginning of the 20th century, it had become increasingly obvious that established geological theories failed to provide a satisfactory explanation for the structure of the surface of the Earth. The traditional assumption that the gradual cooling of the Earth was causing its surface to wrinkle, creating folds that produced mountain ranges and ocean beds, failed to account for the more elaborate deformations found, for example, during an excavation of the Swiss Alps to build the Simplon tunnel between 1898 and 1915.

A new model was called for, and in 1910, although he wasn't the first geologist to propose that the Earth's continents had moved over a long period of time, Alfred Wegener (1880–1930), a German meteorologist and geophysicist, developed the first comprehensive hypothesis of continental drift. His theory was based on the observation that the east coast of South America and west coast of Africa could be fitted together like a giant jigsaw puzzle. He was convinced that they had formed part of a single landmass that had somehow split apart, and looked for evidence to support his idea.

Wegener found that both the geology and the fossils of animal and plant populations up to about 200 million years ago, during an era in geological history known as the Mesozoic, were remarkably similar. Later, the animals and plants appeared to have evolved into separate African and South American species. Wegener argued that this could only have happened if the space left by the separating continents had been filled by a barrier of ocean that prevented the inhabitants on either side from breeding with one another.

Wegener's meteorological interests encouraged him to look for evidence of climatic conditions, too, in the geological past to support his hypothesis. Geological studies showed that glaciation had occurred during the Permian, the period before the beginning of the Mesozoic era, in South America, Africa, India and Australia. This would have been surprising had the continents been located in the same position that they occupied today. It made perfect sense, however, if, as Wegener believed,

Janet Street-Porter

EDITOR, *THE INDEPENDENT ON SUNDAY*

I grew up in part of a terraced house in Fulham, West London, with a small back garden and an outside toilet. My first experiences of nature were both mysterious and frightening. Every time I went to the toilet, a new large spider had spun a web in a corner of it, and I would sit willing this terrifying creature not to notice me during my visit. Each year, my sister and I would go to Penn Ponds in Richmond Park and collect frogspawn in jamjars. Eventually, we would transfer the little tadpoles to a small pool in our garden — but where did they all vanish to? One night I visited the outside loo and discovered a new visitor — a small frog, even more potentially dangerous than the spiders.

I joined the children's club at The Natural History Museum when I was about 8 or 10, and spent many Saturday mornings drawing everything from birds to bears. It was an invaluable haven of peace, and I used to develop my skills at the drawing board, long before I decided to study architecture. At my secondary school I studied biology, and soon solved two of my childhood fears: common spiders can't bite and tadpoles become harmless frogs. I would spend whole mornings in the Museum, opening drawer after drawer of almost identical beetles and marvelling at gaudy moths, pinned out in regimental lines. I would photograph and draw the dinosaurs from every angle, and soon they loomed large in my dreams. I have to admit I was never really interested in the evolution of various species — with me, appearance was everything, and I would scour the Museum for more and more outlandish exhibits.

Pond dipping for jamjar treasures.

As I grew older, I fell in love with the building itself — it was probably one of the reasons I decided to train as an architect. It is surely one of the most spectacular, romantic and captivating places in London, with exotic decoration, grand staircases and a massive sense of scale. I fantasized about holding parties in amongst my friends the sharks (the source of great excitement when the new gallery opened) and the pterodactyls.

To me, a museum of natural history should open your eyes to the world we live in, and be a place you can endlessly revisit to discover new pleasures, fresh nuggets of information about the living world, and new ways of looking at nature. I worry that my way of entering this world, by drawing it, will be replaced by the less creative and challenging exercise of simply logging on via the internet. Seeing and doing for yourself, as in the primary experience of walking round a real exhibit or gallery, is irreplaceable.

I spend a lot of my time walking the countryside, and nothing gives me more pleasure than seeing an unexpected creature cross my path — a fox, badger, stoat or grouse, or even an adder basking in the sun. Our uncultivated moorland is a fragile environment, and hopefully I'm sympathetic to its natural inhabitants. I'm just visiting their home.

J Street-Porter

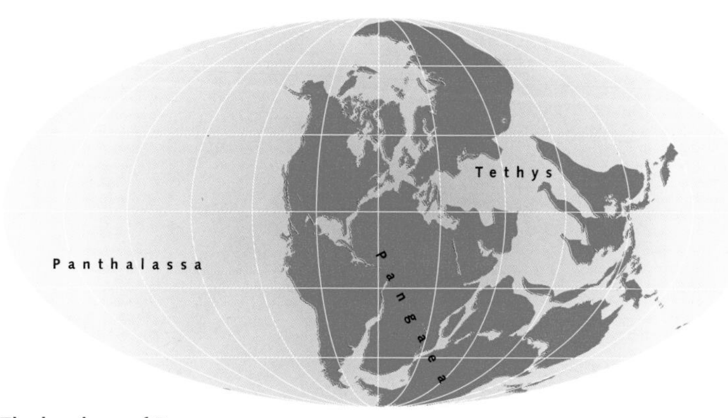

The break up of Pangaea and subsequent movement of the continents, (above) 248 and (below) 2.5 million years ago.

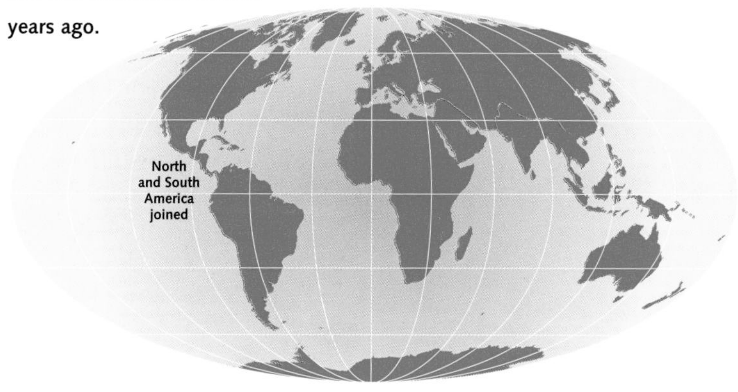

all the continents had once been part of the same landmass, Pangaea — a single super-continent, surrounding the south pole — that had subsequently split apart into sections that gradually moved northwards.

Evidence of the tropical climates once enjoyed by Europe and North America lent further weight to the theory, suggesting that they had gradually moved through a position on the equator. Wegener believed that mountain ranges like those found on the west coast of South and North America could be explained by the crumpling effect of the Earth's crust resulting from the movement of landmasses, while the Himalayas had been created by the collision of India and Asia.

FACING UP TO CRITICISM

Although Wegener had accumulated considerable evidence to support his hypothesis, his idea gained little support. Many geologists, particularly in Britain, criticized his methods, because his arguments were based on facts gathered from a survey of geological literature rather than original observations made in the field. But it was Wegener's failure to suggest a plausible mechanism for continental movement that drew the most criticism.

In the early 1900s, geologists had discovered that the density of rock varied over the world's surface. This suggested that the Earth's crust consisted of two layers, with the continents formed from sial, the lighter layer of rock, resting on top of sima, the denser layer below. Wegener suggested that the top, sial, layer was 'floating' on the static sima below and that either centrifugal or tidal forces accounted for its movement. His mechanism was criticized because it seemed unlikely that these forces were powerful enough to overcome friction between the layers of rock.

Without further evidence, Wegener's model seemed no more plausable than any other hypothesis at the time, including the popular suggestion that the continents may have been separated by a bridge of land that sank beneath the surface of the sea.

A NEW ANGLE

In 1924, the British geologist Arthur Holmes (1890–1965) came up with an exciting new explanation for continental drift. He suggested that the continents were carried over the surface of the Earth by convection currents of molten rock. Acting like conveyor belts, these currents rose from the mantle, the deepest layer of the Earth, and spread horizontally over its surface,

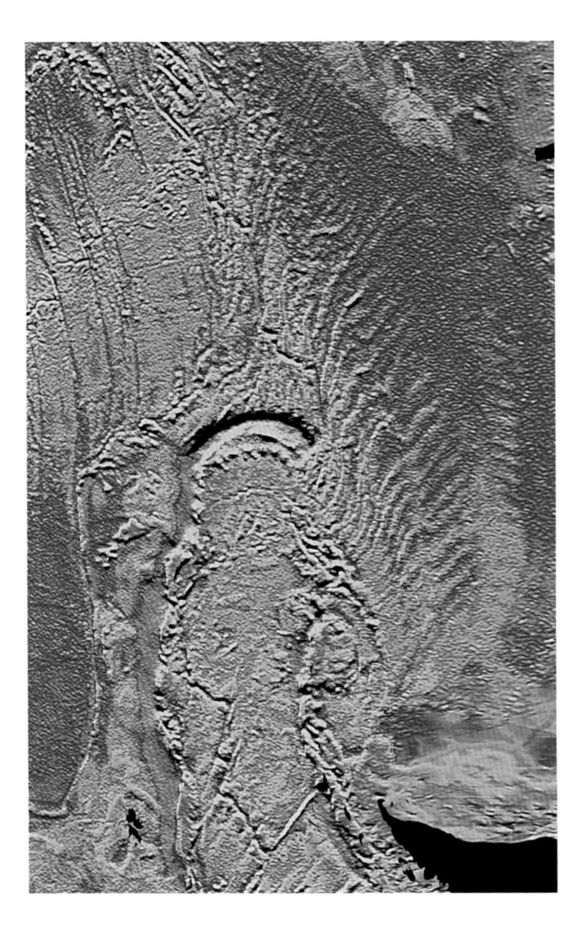

LEFT: **This satellite image of the South Atlantic seabed shows the San Andreas Fault in California — the notorious earthquake region in California, produced by the strike slip boundary between the Pacific and the North American plates.**

RIGHT: **San Andreas Fault.**

before cooling and sinking back downwards. Like Wegener's, this model was largely speculative and, although it seemed reasonable, Holmes was the first to recognize that 'ideas of this kind, specially invented to match the requirements, can have no scientific value until they acquire support from independent evidence'. It took another forty years before the 'independent evidence' described by Holmes started to emerge from surveys of the ocean bed and paleomagnetism, the study of the Earth's ancient magnetic field.

STRATEGIC ADVANTAGE

Governments seeking military advantage invested heavily in oceanography during the Second World War and the Cold War years of the 1950s and 1960s. Understanding and mapping the magnetization of the sea floor was particularly important to the US Navy, who believed it would help them to improve detection of enemy submarines. Mapping the ocean floors produced remarkable results that turned traditional geological expectations on their head. It showed that the ocean crust was comprised of rock far younger than that of the continental crust, and suggested that some of the most significant geological activity was taking place in the ocean beds rather than on land.

Geologists knew that the seabed contained ridges and deformations but, for the first time, this intensive survey generated a map on a global scale. The map showed that ocean trenches and chains of mountains similar to the Mid-Atlantic Ridge, the one known to run the length of the North and South Atlantic under the sea, were also present in other oceans. In the 1960s, American geologist Harry Hess (1906–69) suggested that these ridges were where molten rock rose from the mantle to form new sea floor, and ocean trenches marked the regions where old ocean crust was being pulled, or subducted, back down into the Earth.

Hess's model gained support in 1963, when English geophysicists Fred Vine (1939–88) and Drummond Mathews (1931–97), and independently, Canadian geologist Lawrence Morley, recognized that the curious magnetic patterns produced by the rocks surrounding mid-ocean ridges corresponded with Hess's model of 'sea-floor spreading'. Paleomagnetism generated the 'independent evidence' that encouraged the majority of the geological community to begin to accept the plausibility of continental drift.

All that remained was for a number of geophysicists to take the concepts of sea floor spreading and continental drift and develop them into the more general theory of plate tectonics and, by the end of the 1960s, the conceptual revolution that Wegener had begun in 1910 was complete. Incorporating the ideas of Canadian geophysicist John Tuzo Wilson (1908–93) about the movement of the Earth's crust, the theory suggests that the outer layer is divided into rigid segments, or 'plates' that slip over the partially molten layer below — and that it is the motion and interaction of these plates that explain phenomena like volcanism, mountain building, deep sea trenches, earthquakes and ocean basins. Wegener may not have been able to predict the precise mechanism by which the continents moved, but the plate tectonic theory shows that his general model of particular continental movements was surprisingly accurate.

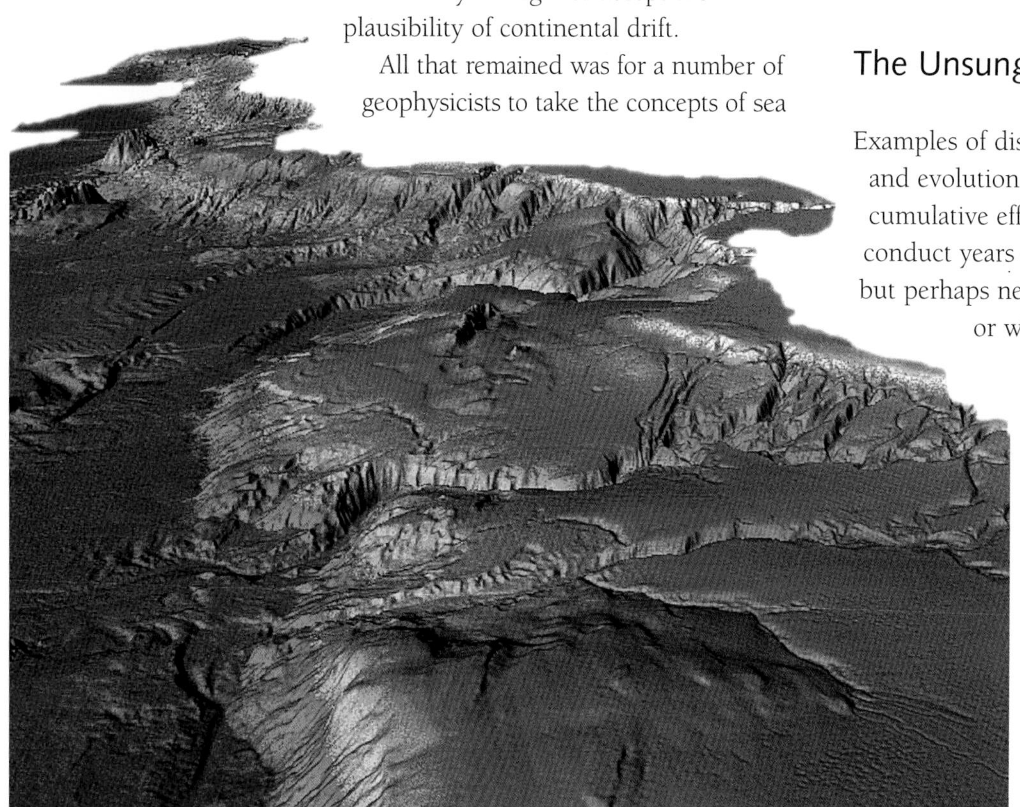

Sonar image of the ocean floor in California.

The Unsung Heroes

Examples of discoveries like continental drift and evolution show that progress is the cumulative effort of many scientists, who conduct years of critically important work but perhaps never make a major breakthrough, or who do but fail to gain the proper acknowledgement for their achievement.

KEPT OUT OF THE LIMELIGHT

Take the case of British physical chemist and X-ray crystallographer Rosalind Franklin (1920–58), for example. In the late 1940s and early 1950s, she perfected the technique of developing X-ray crystallographic images

LEFT: **Rosalind Franklin.**

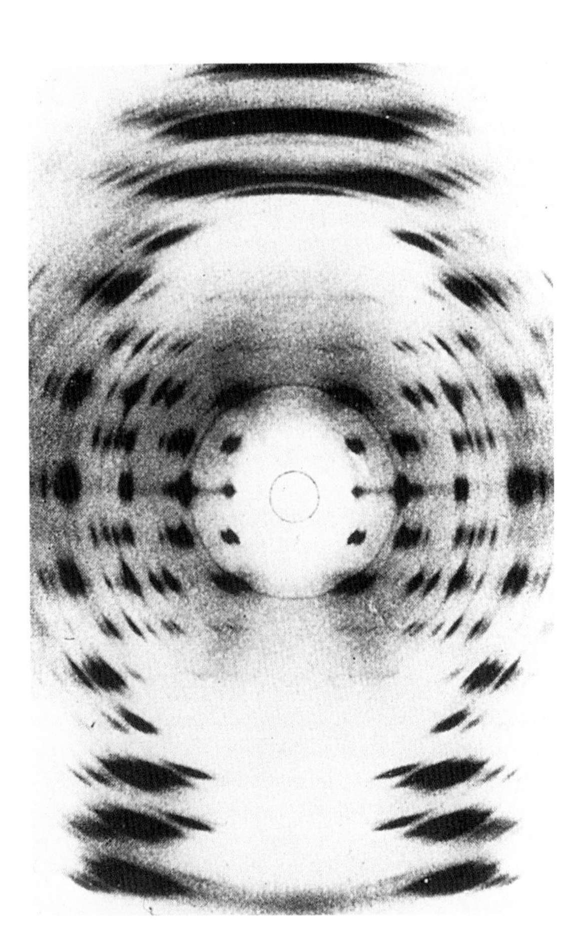

RIGHT: **An X-ray crystallograph of DNA prepared by Rosalind Franklin.**

of DNA. It was one of Franklin's images that inspired James Watson to first think that DNA might be shaped like a double helix.

Franklin's technical skills provided vital evidence to support Crick and Watson's ideas but she was given little acknowledgement in Watson's personal account of the discovery, *The Double Helix*. In Lawrence Bragg's foreword her supervisor at King's College, Maurice Wilkins, was acknowledged, for his 'long, patient investigation', as was Crick and Watson's 'brilliant, rapid final solution', but there was no mention of Franklin's contribution. Watson's account of her role within the main body of the text is highly critical, and at times verges on a character assassination:

'So it was quite easy to imagine her the product of an unsatisfied mother who unduly stressed the desirability of professional careers that could save bright girls from marriage to dull men.'

It was only after she'd died that Watson acknowledged his unfair treatment by adding an epilogue to his book, in which he wrote: 'In 1958 Rosalind Franklin died at the early age of thirty-seven. Since my initial impressions of her, both scientific and personal (as recorded in the early pages of this book), were often wrong, I want to say something about her achievements. The X-Ray work she did at King's is increasingly regarded as superb.'

Watson goes on to praise the quality of her work in full, 'realizing years too late the struggles that the intelligent woman faces to be accepted by a scientific world which often regards women as mere diversions from serious thinking.'

Why Scientists are not Gods

The great American popularizer of science Isaac Asimov once wrote that Imhotep, a famous Egyptian priest and physician from the 27th century BC, was the only scientist ever to be recognized as a god. Some might quibble over Asimov's classification of a scientist, but his tongue-in cheek observation raises an interesting point.

Scientists don't have all the answers; they can only provide the most reasonable explanations of natural phenomena based on the evidence that is available at any given time — and, as we have seen earlier in this chapter, everything is subject to change. As new evidence comes to light, even the most established theories can be superseded so that science can progress. Although scientists endeavour to be as objective as possible, their interpretation of facts is inevitably shaped by their private beliefs and experiences. This explains why different experts can present conflicting views, from the same body of evidence, on issues such as the danger of contracting BSE from eating beef on the bone, the effects of environmental pollutants on our planet, and whether or not there is life on Mars.

Scientific research, as the case of Rosalind Franklin demonstrates, can be extremely competitive and, occasionally, as we saw with the theory of divine creation in the 19th century, theories persist more because of the power enjoyed by the scientists who propose them than for any objective reason such as accuracy. Fortunately, the communal nature of science, based as it is on the co-operation and consensus of scientists working on the same problems around the world, ensures that there are enough independent checks and balances in place to ensure that the better explanation will eventually prevail.

Eygyptian bronze statue of the god Imhotep.

The Big Picture

'Nothing in biology makes sense except in the light of evolution.'

Theodosius Dobzhansky (1900–1975)

These large mammals may appear to have little in common but a closer inspection reveals that whales, elephants and rhinoceroses aren't so very different after all.

On first encounter, a natural history museum can seem like a mishmash of curiosities. What can possibly be the relationship between dinosaurs, dandelions, meteorites, and a startling Victorian display case of miniature birds? It's the same dilemma many would face if asked to explain natural science to aliens from another planet.

What would most people start to describe? Plants and animals? The dinosaurs? Conservation? Ecology? If we feel confused, it's because the scope of natural science is vast. It not only extends across the scale from the submicroscopic and microscopic world of molecules and cells to the global distribution of biological communities, but also includes a history that stretches back over billions of years, to the moment the Earth began. So how *do* we impose order on this dizzying array of information? What would help us to stand far back enough from the detail to see, in the broadest of terms, how the various 'pieces' of the natural science 'jigsaw' fit together?

Dividing the study of nature into different disciplines, such as botany and mineralogy, or into areas of study like genetics and taxonomy, creates artificial categories that make this information more manageable. But it is also partly responsible for creating the misleading impression that these areas are somehow separate and unrelated. In reality, the majority of natural scientists believe the opposite is true:

that all living organisms, in their rich diversity, share common properties, and can be traced back over billions of years to a single ancestor — the result of a continual process of transformation in response to the gradual and catastrophic changes that have taken place on Earth since its formation.

Evolution is the core theme that ties together our understanding of the natural world. One of the greatest biologists of the 20th century, Theodosius Dobzhansky, once said that 'Nothing in biology makes sense except in the light of evolution'. Extend biology to include the whole of natural science and this powerful statement remains equally true. Evolution can also be divided into a number of subthemes, which explain why specimens like the towering skeleton of *Diplodocus carnegiei* aren't after all so very different from the rich diversity of specimens on exhibit in natural history museums. Browse through the range of examples in this chapter to grasp these key ideas, before you take a more detailed look at the scientific background in the following chapters.

What is the relationship between *Diplodocus* and *Triceratops*, a meteorite and a dandelion?

Understanding the Earth

A DYNAMIC PLANET

In 1995, lava flows and ash produced by a volcanic eruption on the Caribbean island of Montserrat forced people on the southern side of the island to flee their homes, and destroyed the capital city of Plymouth. In June 1998, an earthquake measuring 6.9 on the Richter scale triggered mudslides, and destroyed up to 78 villages in Takhar and Badakhshan in north-west Afghanistan. It occurred a year after an equally devastating earthquake killed approximately 4000 people in the same region. In February 1999, a series of avalanches in the Austrian Alps swept through ski resorts and killed 38 people.

Horrific tragedies like these that result from volcanic eruptions, earthquakes, tidal waves, avalanches, floods and mudslides provide a sobering reminder that the Earth is a dynamic, constantly changing planet. But these cataclysmic events aren't the only things that alter the landscape. There is continuous, extremely gradual alteration taking place

Ash covers Plymouth, the capital of Montserrat after the 1995 volcanic eruption.

all about us, too. Nearly every place on the Earth's surface is rising upwards, moving sideways, tilting, sliding or being eroded by some natural process. Scandinavia, for example, is rising at the rate of 1 cm per year, literally bouncing back from the weight of an immense ice sheet, 2–3 km thick, that covered it 40,000 years ago.

HOW THE EARTH WAS SHAPED

The processes that shape the Earth today are the same as the ones that have operated since its origins. But it is only in the last two hundred years that scientists have begun to understand the processes that drive these changes and appreciate the sheer enormity of the timescales involved.

Most scientists agree that the solar system was created from a nebula (a massive cloud of dust and gases) that circled the Sun nearly five billion years ago. Although there are a number of different hypotheses to explain the origins of the Earth within this system, there is a general consensus that it is about 4.6 billion years old. The newly formed Earth continued to gather up asteroids and comets in its path until, after several million years, it reached its present size. Scientists think that the new planet started out as a homogeneous mixture of materials but gradually settled out into a number of different zones, or 'onion-skin layers'. Gravity drew the heaviest materials, such as iron, to the centre, while the lighter materials, such as magnesium and iron silicates, rose to the surface to form a primitive crust, and the gases escaped to form a primitive atmosphere. During this differentiation process, the temperature of the Earth soared, due to the combined effect of the internal heat released from the decay of radioactive elements within its interior and the constant bombardment of meteorites on its surface.

A VERY DIFFERENT WORLD

The surface of the newly formed Earth was very different from the one we see today. In the beginning, there were no atmosphere, continents, or oceans, and scientists believe they were created over billions of years, the by-products of the gradual settling and differentiation of materials within the Earth's interior. One popular theory suggests that the intense heat generated from this process produced an ocean of molten rock and convection currents that transported some of this heat to the surface. These currents cooled the Earth and solidified its crust, while the iron core remained molten and still remains so to this day.

Scientists believe that the heat generated by the differentiation of materials within the Earth's interior triggered volcanic eruptions that released oxygen and hydrogen from compounds such as potassium aluminium mica. The oxygen and hydrogen combined to form water vapour. This, together with a number of other gases,

HOW DO WE KNOW THE AGE OF THE EARTH?

STRATIGRAPHY: RELATIVE DATING

Until the early 20th century, geologists had no direct way of calculating the absolute age of the Earth, but they were able to reconstruct a relative time scale by studying the sequences of sedimentary rock layers in different parts of the world. Sedimentary rock is formed when layers of sediments, such as sand, mud and silt, deposited in oceans, lakes and rivers, are compressed over millions of years. Gradually, new layers of sediment are added, forming a stack or column, with the oldest layers of rock at the bottom and the youngest layers at the top.

Different parts of the sedimentary rock throughout the world chart different eras of geological time. Geologists gradually pieced together the various sequences to reconstruct a single sequence that charts the chronological order of past events, from the formation of the Earth to modern times. Names of time intervals, such as 'the Cretaceous' and 'the Jurassic', were allotted to each period of geological time before anyone knew how long each period had lasted, or how long ago it had occurred in relation to modern times.

Identifying each layer of rock is not always an easy process. One of the difficulties is that natural processes have eroded and deformed the Earth's crust, overturning sedimentary layers in many places, so that they do not necessarily appear in the sequence in which they were formed. At first, geologists thought they could look at the minerals in each

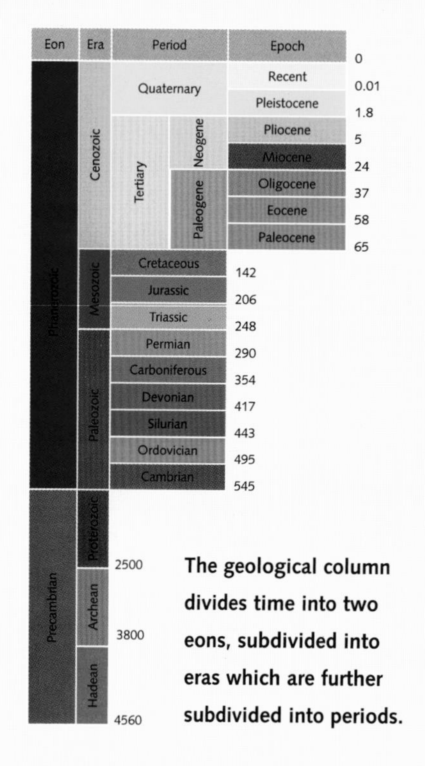

The geological column divides time into two eons, subdivided into eons, subdivided into eras which are further subdivided into periods.

layer to establish the correct sequence of formation, but detailed surveys showed that the mineralogy of rocks wasn't necessarily unique to any particular geological period. The researchers found that the same type of rock formed at different times throughout the Earth's history, while different types of rocks were just as likely to form in different locations at the same time.

CLOCKS IN ROCKS: ABSOLUTE DATING

In 1905, New Zealand physicist Ernest Rutherford (1871–1931) was the first to suggest that minerals containing radioactive forms, or isotopes, of elements could be used in dating rocks (geochronology). Along with other pioneers of nuclear physics, Rutherford observed that radioactive isotopes disintegrate spontaneously to form more stable 'daughter' isotopes, and that this process of disintegration, called radioactive decay, takes place at a constant rate, like the regular beat of a clock marking time. The 'geological clock' in a rock starts when the rock is first formed and unstable isotopes within its crystallized minerals begin to decay. By comparing the proportion of radioactive isotopes to the proportion of stable 'daughter' isotopes that remain in the rock, scientists can calculate the rock's absolute age in years. Uranium-lead, potassium-argon and carbon-14 dating are all forms of absolute dating based on the principle of radioactive decay.

This breakthrough in absolute dating, and the development of increasingly sophisticated methods to calculate the age of ancient rocks, encouraged geochronologists to turn their attention to the age of meteorites, where they made a startling discovery. Gradually, as more and more meteorites were tested, it became apparent that they were all approximately 4.5 billion years old, regardless of their composition, or the date on which they collided with the Earth. Since no older dates have been obtained from Earth or Moon rocks, and since we believe that the Sun, planets and meteors have all been formed at roughly the same time, this date — 4.5 billion years old — is regarded as the most likely age of our Solar System.

including nitrogen, carbon dioxide, carbon monoxide, hydrogen, hydrogen chloride and methane, formed a primitive atmosphere, and eventually oceans, over hundreds of millions of years.

At the same time, the primeval crust continued to melt and solidify over and over again, forming rocks that were broken down by rain and other erosive agents. The residue produced by this weathering process formed sediments. These were either resorbed into the Earth's crust or transformed into new rocks by the gases and heat that continued to escape from its interior. Experts calculate it would have taken approximately 2.6 billion years for the continents to form as a result of these processes after the differentiation of the Earth's mantle and core.

PLATE TECTONICS

Many of the major geological features that have characterized the Earth since the formation of its primitive continents, oceans and atmosphere can be explained under the unifying theory of plate tectonics. Briefly, the theory suggests that the outermost layer of the Earth, the lithosphere, is divided into approximately twelve rigid plates that sit on a partially molten layer of rock, the asthenosphere. Convection currents, driven by the internal heat produced by radioactive decay within the Earth's interior, circulate through the asthenosphere, causing the very gradual movement of these plates, and many large-scale geological phenomena, including earthquakes, volcanic eruptions, mountain-building, and the formation of ocean basins and deep-sea trenches, occur at the boundaries between them.

This section began with examples of the catastrophic and gradual changes that continue to alter the Earth's surface. These changes are produced partly as a result of plate tectonics and the geological processes that occur within the Earth's interior but, as we will see in later sections, they are also produced by processes that occur on the Earth's surface. Understanding how these opposing processes operate and interact with one another is the key to understanding the Earth today, how it evolved over geological time, and — perhaps most exciting of all — how it came to support life.

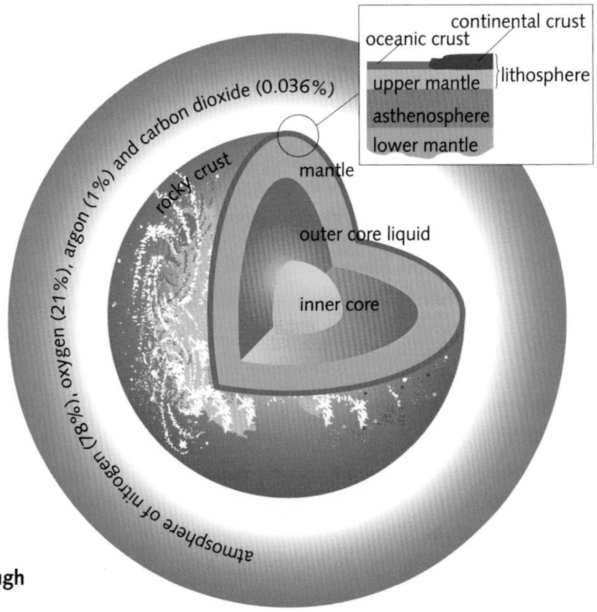

Cross-section through the modern Earth.

Internal Processes

The Earth's interior behaves like a gigantic, pressurized heat engine that generates enough energy to melt rocks and drive the processes that move continental plates, build mountains, and produce earthquakes and volcanic eruptions. Understanding how these processes operate is an important part of extending our understanding of what the Earth is and how it works, but how do scientists investigate this impenetrable region of our planet? Mark Welch, a mineralogist, explains:

HYDROTHERMAL VENTS

In 1977, geologists began a series of dives around the Galapagos Islands, using a deep-sea submersible. Their aim was to locate cracks in the ocean floor that, according to the theory of plate tectonics, marked the point at which new ocean crust was being formed. Theoretical studies also predicted cooling of the newly formed crust by circulating seawater, and the researchers hoped they might find sites where superheated water was being vented. The expedition was a spectacular success. The scientists not only found black smokers — hydrothermal vents discharging black plumes of superheated water, but also discovered completely unexpected communities of exotic marine organisms flourishing.

At first glance, these animals looked like known species of mussels, crabs, clams and tubeworms, but a more detailed inspection revealed that some were very different indeed. Here were complete communities that relied not on sunlight and photosynthesis for their primary source of energy, but on hydrogen sulphide produced from deep within the Earth's crust. The fluid from hydrothermal vents contains a rich supply of hydrogen sulphide, which combines with trace metals to precipitate out as insoluble metal sulphides. These build up to form the chimney-like structures that give black smokers their name.

Despite the fact that they are found in some of the most inhospitable and inaccessible regions on Earth, hydrothermal vents continue to draw intense scientific interest. Studying them also provides vital clues as to how life began. Genetic studies, for example, have revealed that the hydrogen-sulphide-eating bacteria, which flourish in scalding temperatures in and around hydrothermal vents, are the closest living descendants of the universal ancestor from which all living organisms originated, over four billion years ago. This new evidence challenges the long-held belief that life began on the surface of the Earth, suggesting instead that it may have its origins in the pressure-cooker conditions deep within our planet.

The prohibitive cost of deep-sea expeditions makes investigations of hydrothermal vents extremely rare events. But some scientists, like Eva Valsami-Jones have found other, less costly, ways of unravelling the mysteries of hydrothermal vents:

'We have discovered a vent site in relatively shallow waters near the Greek island of Milos that we can access without the use of expensive equipment. These active, shallow-water vents behave in the same way as those found in the deep sea, and we are hoping that understanding the changing chemistry of the hydrothermal waters near Milos will give us a better understanding of the processes that occur in other less accessible regions on Earth. Scientists can learn a lot about surface processes using sophisticated modelling techniques, but nothing quite matches the satisfaction of working directly in the field. The unique vent site near Milos is a perfect natural laboratory, where we can monitor chemical changes and observe the effect of fluctuations within the environment at first hand.'

LEFT: **Shallow hydrothermal venting in Milos.**

RIGHT: **The bay in Milos where scientists are investigating hydrothermal vents.**

'Major advances in understanding geophysical and geochemical processes occurring deep within the Earth's crust and mantle have come from laboratory studies that simulate the very high-pressure / high-temperature conditions of these deep regions of the planet. While the early laboratory studies of the 1940s and 1950s could reach modest pressures, equivalent to depths of up to 70–80 km, recent technological advances have allowed processes occurring at depths of up to 200 km to be investigated routinely. More sophisticated, non-standard apparatus now allows access to pressures corresponding to depths of 600 km.

Scientists developed ideas about the likely composition and mineralogy of the Earth's upper mantle from laboratory experiments on rock compositions corresponding to those of exotic rock fragments contained in lavas, most notably the rock-type known as peridotite. These studies revealed that some of the mineral assemblages of peridotite (pyrope garnet; olivine and pyroxenes in the case of garnet-peridotite) are stable only at depths of more than 60 km, so that they are likely to have originated from the mantle lying below the crust.

These rock fragments are preserved without reacting to low-pressure (pyrope-free) assemblages because they were carried from the mantle to the surface by magma (molten rock) originating in the mantle. Peridotite is one of the commonest rock-types to occur as fragments in basaltic lavas, suggesting that it is an important mantle constituent. Further support for this idea came from the recognition that large bodies of rock emplaced into the continental crust along major thrust faults represent cross-sections through the oceanic crust and upper mantle, with the mantle rocks dominated by peridotite.

Garnet-peridotite, a major constituent of the Earth's upper mantle. Volcanic rock (grey matrix) containing fragments of peridotite (green rock) from the Earth's upper mantle. The peridotite here is composed mainly of olivine and pyroxenes. The largest fragment is about 10 cm across.

Laboratory simulation of conditions within the Earth's interior is helping Earth scientists to understand and predict how eruptions behave and the mixture of materials they produce. Early work focused on the changing chemical composition of magmas and lavas within volcanic systems. For example, high-pressure / high-temperature experiments on peridotite showed that varying degrees of melting of this type of rock could produce most of the compositional varieties of the commonest lava type, basalt. The fluidity of magma changes with its chemical composition and is a key factor in determining how an eruption will behave, for example whether gases can escape without causing an explosion or generate billowing aerosols of lava dust, ash and gases that travel around the world.

Much subsequent high-pressure experimental research has been concerned with understanding how high pressures affect the minerals (olivine, pyroxenes and garnet) making up peridotite. Laboratory experiments have revealed that olivine, the main constituent of peridotite, undergoes a series of dramatic structural transformations at pressures corresponding to those at the depths of very deep earthquakes that occur at destructive plate boundaries ('subduction zones'). Earth scientists have used this information to construct geophysical models that explore the interactions between deep- and shallow-focus earthquakes.

Another area of topical interest is the behaviour of water-bearing minerals in subduction-zones, the main regions where water is recirculated within the Earth's mantle. Water has a major effect upon how rocks melt and deform, and is, in part, responsible for the magmatism that created the Andes and the volcanic island chains rimming the Pacific Ocean. Scientists estimate that about 90% of the water entering the mantle in subduction zones is held in hydrous (water-containing) minerals of the oceanic crust. The depths at which these minerals break down and lose their water depend upon the thermal gradients in subduction zones, which can be very low. Laboratory experiments simulating subduction-zone conditions reveal that in younger, hotter subduction zones, high thermal gradients result in hydrous minerals breaking down at depths of less than 250 km, releasing water into the overlying mantle where it can cause extensive melting. In old (over 50 million years) cold subduction zones, experiments suggest that hydrous minerals persist and transport their water to much greater depths (below 500 km). Understanding the geological consequences of this very deep source of water is an area of intense current research, for example in understanding the causes of very deep-focus earthquakes.'

On the Surface

The Earth's internal heat drives the geological processes that build mountains and continents. But it is a combination of the Sun's radiant energy and gravity, which draws water, ice, rocks and soil down slopes, leaving a trail of destruction behind them, that drives the external processes that attack and erode them. To an Earth scientist, the Earth's surface contains much more than the rock, soil and minerals that form its outermost layer of crust. It also includes the protective atmospheric gases that surround it, the water that circulates over and through it, and the rich variety of organisms that live on it. None of these systems operates in isolation, and scientists are only just beginning to make sense of the complex ways in which they interact with one another, to literally change the face of the world.

The Andes, Torres del Paine, Chile.

Glencoe, Scotland.

THE EARTH'S CRUST

It might look solid, but the Earth's crust is in a perpetual state of animation, bound in a cycle of renewal and decay. Molten magma, forced upwards to the surface through the action of plate tectonics, crystallizes to form new rock, only to be eroded by natural agents, such as wind and water, and transported to new destinations to form new rock. In the right conditions, these transported sediments can accumulate, become compacted and form solid rock before the cycle of erosion and transportation begins again. This is how sandstone is formed from sand, and shale from mud and clay. The formation of sedimentary rock can take a very long time. Some sediments, for example, have remained loose for 30 million years. We can see evidence of erosive change all round us, from the facades of stone buildings, dissolved by the acid by-products of industrial pollution, to dramatic landscapes like Glencoe, in Scotland, literally carved from granite rock by the movement of ice over the last two million years.

The 'Sunday Stone' specimen on display in The Natural History Museum, London provides an unusual example of surface change. The pattern of black-and-white bands in the sedimentary rock represents a natural 'calendar' of weekly activity in a Tyneside coal mine in the 1800s. The sediment was formed from layers of barium sulphate crystals settling out in a water pipe within the mine. The black bands are the result of coal dust darkening these white crystals when the mine was in operation, while the unmarked white bands of barium sulphate were formed during periods of inactivity on Sundays and public holidays.

Earth from space.

THE HYDROSPHERE

The familiar photographs of Earth taken by space satellites explain why our world has earned itself the nickname 'the blue planet'. Predominantly blue, the satellite pictures reveal vast expanses of oceans veiled by swirling clouds of water vapour. Over 70% of the Earth is covered with approximately 1.3 billion cubic kilometres of water. No less than 97.3% of the Earth's surface water is held in the oceans, only 2.1% in ice sheets and glaciers and a mere 0.61%

The Sunday Stone.

Dundonell River, Scotland, in full spate.

in groundwater, freshwater streams, rivers and lakes. Only a minute quantity, about 0.001%, exists as vapour in the atmosphere, but it still accounts for a global rainfall of some 4000 million tonnes of water every year. Water circulating between the ocean, land and atmosphere is one of the key agents of surface change. It erodes rocks, transports and deposits minerals, and influences the climate and the weather by storing and redistributing solar energy from the equator to the poles.

The atmosphere protects us from the Sun's harmful radiation.

The Atmosphere

The atmosphere is the thin blanket of gases, held fast to the Earth by its gravitational pull, that makes our planet unique within the solar system. Consisting of mostly oxygen and nitrogen, these gases are not only essential for life, but also act like a shield protecting us from the harmful aspects of the Sun's radiation and, together with the oceans, store and distribute energy, moderating what would otherwise be a far more extreme global climate.

The atmosphere is divided into several layers. About 99% of its gases are concentrated in the troposphere, which includes nearly all the water vapour and clouds, and this is where most of the weather, or atmospheric turbulence, occurs.

The carbon dioxide and water in the atmosphere play a vital role in controlling global temperature, by trapping the heat that radiates back from the surface of the Earth (the greenhouse effect). Without this essential barrier, the heat loss would be rapid and, when not being warmed by the Sun, the Earth would freeze below temperatures that could sustain life. Temperatures as low as minus 120°C, for example, have been recorded on the side of the moon not heated by the Sun.

Ever since humans began burning large quantities of fossil fuels during the industrial revolution, the levels of carbon dioxide (CO_2) entering the atmosphere have continued to increase. In the 1990s, for example, it is estimated that more than five billion tonnes of

If global sea level rose just tens of centimetres, cities like London, and New York, shown here, would probably have serious flooding problems.

CO_2 were being added to the atmosphere each year. No-one knows for sure how this build up of CO_2 will affect the Earth, but experts are concerned that it may be linked to a noticeable increase in global temperatures. Why should they be concerned?

Climatologists believe that an average rise in global temperature of only a few degrees Celsius could melt the ice caps in the polar regions, and cause thermal expansion of the oceans (as the oceans warm they become less dense and expand), resulting in a rise in global sea levels. Estimates vary, but some experts have predicted that sea levels could rise anywhere between 20 cm and 70 cm by the year 2070.

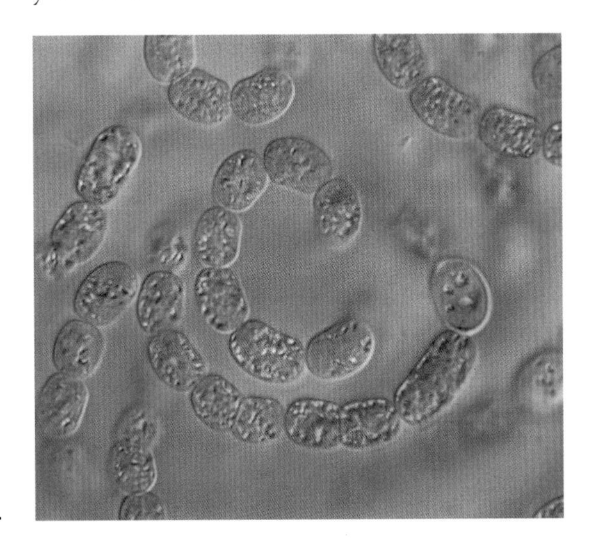

Cyanobacteria.

The Biosphere

There are approximately 6400 km between the centre of the Earth and its surface and another 1000 km beyond this to the outer limits of its atmosphere. Although multicellular organisms have been found at the bottom of 11,000 m deep ocean trenches where the average depth of the sea is 4–5000 m, so far the records for the vertical limits of life are held by bacteria found living even farther down, 750 m beneath the Earth's crust. At the other end of the scale, bar-headed geese have been spotted flying nearly 9000 m above sea level by climbers on Mount Everest, and a Rüppell's griffon vulture collided with an aircraft flying at an altitude of 11,277 m over the Ivory Coast, West Africa, in 1973. The great majority of plants and animals, however, occupy a much smaller range than these incredible extremes, and are concentrated on the land surface and in the upper 150 m of the ocean.

The biosphere includes every region of the Earth that sustains life. Living organisms play a key role in the global cycling of energy and materials, and are responsible for creating the conditions necessary for their own survival. It was the evolution of plants and other oxygen producers, such as cyanobacteria, for example, that is responsible for generating and maintaining the levels of free oxygen in the atmosphere on which most organisms depend.

Themes for Life

Living organisms can be subdivided into a hierarchy of organized components or structural layers that build upon, and co-operate with, one another to sustain life. Atoms, for example, the most basic chemical building blocks, link together to make more complex molecules

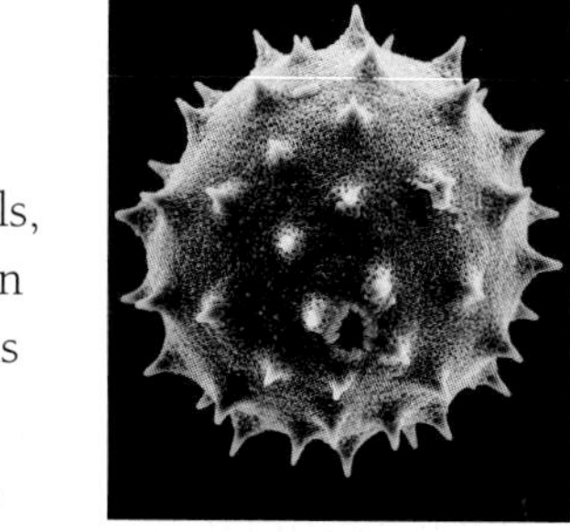
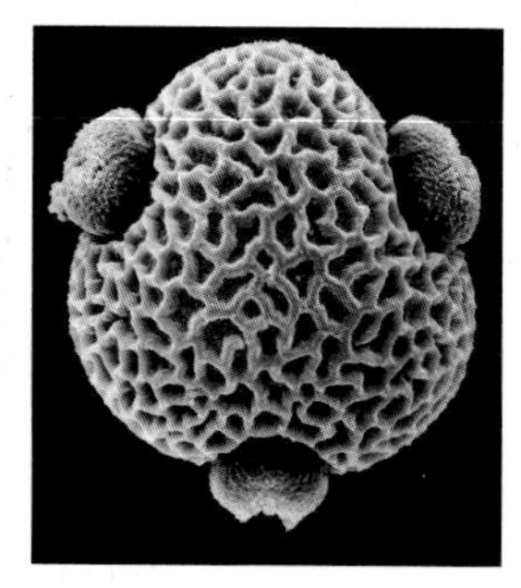
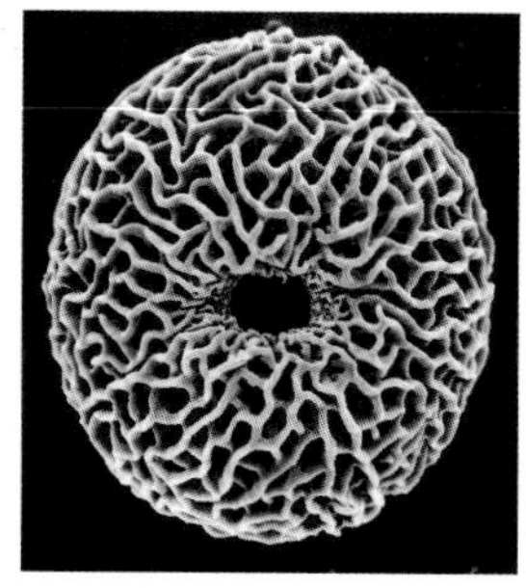

Natural order and complexity extends to the submicroscopic world of individual grains of pollen.

like proteins, sugars and DNA. These complex molecules form the basis of cells, which occur singly in unicellular organisms or in groups, as specialized tissue, in most multicellular organisms. Tissues form the basis of organ systems (the excretory system in animals or the root system in plants, for example), which together form the complete organism. The hierarchy of biological organization extends beyond the individual, too. It extends to populations (groups of the same species living together in one location), and communities (populations of different species that share the same habitat), and culminates in ecosystems, which include the complex interactions of communities with their environment.

You can see natural order at every level of the scale, from the complex structure of

MICROBES, MINERALS AND THE ENVIRONMENT

The long term ecological effects of many pollutants, such as the inorganic radioactive chemicals released into the atmosphere as a result of the nuclear reactor failure in Chernobyl, are not well understood. However, many pollutants, whether of atmospheric or terrestrial origin, will ultimately bind to the surface of minerals found in soils and sediments. Mineralogists are hoping to learn more about the environmental fate of these contaminated materials and possible methods for treating them, by studying the ways in which microorganisms interact with them and mobilize the noxious chemicals at their surface.

Microorganisms sit at the top (as consumers and decomposers), and bottom (as primary producers) of almost all food chains, and therefore play a major role in the transformation of contaminants at the Earth's surface. There are many ways in which microbes can react with pollutants held at mineral surfaces. They can react physically, absorbing toxic chemicals, for example, or bind with them via specialized structures in their membranes or cell walls. They can also act as catalysts, making possible or speeding up chemical reactions that can either increase or reduce the mobility of pollutants — and, therefore, the potential uptake of these same contaminants by plants and other organisms higher up the food chain. Many scientists hope that by studying the ways in which certain microbes can immobilize toxic pollutants, or transform organic contaminants, such as pesticides, into harmless compounds, they will be able to propose effective new methods for treating contaminated waste.

Chemical waste dump, Nandesari, in Gujarat, India.

cells and individual grains of pollen to the regular patterns formed by flocks of migrating birds. The TEM (transmission electron microscope) image of the crystal structure of bone (right) illustrates this point perfectly, demonstrating that at this most fundamental level living organisms are built from minerals, the same inorganic building blocks from which the Earth is formed.

The crystals in human bone are so small that a TEM must be used to see them. They are visible as regular rod shapes, and the bands or dots in them show the position of individual atoms.

MORE THAN THE SUM OF THE PARTS

All organisms, even the tiniest single-celled microbes, are astonishingly complex, yet it isn't so much their components as the way in which these parts interact with one another that produces the characteristics we associate with life. Moving upwards through the hierarchy of biological order, interaction leads not only to increasing levels of complexity, but also generates new properties that aren't apparent at simpler levels of organization. By themselves, for example, none of the individual components of the human circulatory system — the arteries, veins, capillaries and the heart — can distribute blood around the body.

Interaction between individuals within a population can be every bit as important for survival as the complex internal interactions within living organisms, as studies of honeybees have shown. Building on much earlier research, videotape studies in America reveal that a swarm of honeybees operates as a single collective 'brain' when selecting a new home. On returning to the hive, scouting bees perform dances to communicate to the other bees the location and suitability of possible sites; these, in turn, are replicated by individual bees within the swarm.

At first, several separate dances are apparent but, gradually, a consensus is reached as all but one of the scouts gives up. A single dance signifies a final decision, and the swarm moves off to settle down in its new location. Co-operation enables the group to choose the most suitable home, even though individually the bees may not have been in contact with all of the scouts and therefore may not have all of the information about all of the sites.

Bee swarm.

MADE OF CELLS

Cells represent the most fundamental level of biological structure capable of sustaining all the activities associated with life. Some organisms, such as bacteria, among the smallest but most abundant forms of life on Earth, consist of only one cell. At the other end of the scale, large plants and animals consist of millions, sometimes billions, of specialized cells that perform different tasks. Specialized cells such as blood cells and skin cells in animals, or root-tip cells and vascular cells in plants, can look very different, and may have different life spans, but they all share the same basic properties.

THE THREE DOMAINS OF LIFE

Until the late 1970s, the classification of the natural world was divided into two domains, and organisms were described as either prokaryotes (bacteria and cyanobacteria) or eukaryotes (protists, plants, fungi and animals), depending on the structure of their cells. Most scientists assumed that prokaryotes were the most primitive, and had descended from a common ancestor several billion years before the appearance of the first eukaryotes, around a billion years ago. But in the late 1970s, American microbiologist Carl Woese and his co-workers made a stunning discovery. Using the latest genetic sequencing techniques, they discovered that a group of organisms called archaebacteria, previously thought to be a peculiar strain of bacteria, represented an entirely new domain of life. Structurally, they may look very similar to bacteria, but comparisons of the gene sequences from representatives of each domain show that archaebacteria are as different from bacteria

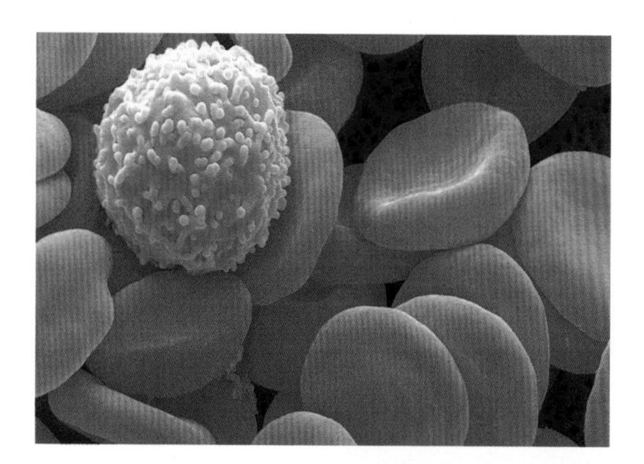

LEFT: **Red and white human blood cells seen through an electron microscope.**
RIGHT: **Human epithelial skin cells seen through an electron microscope.**

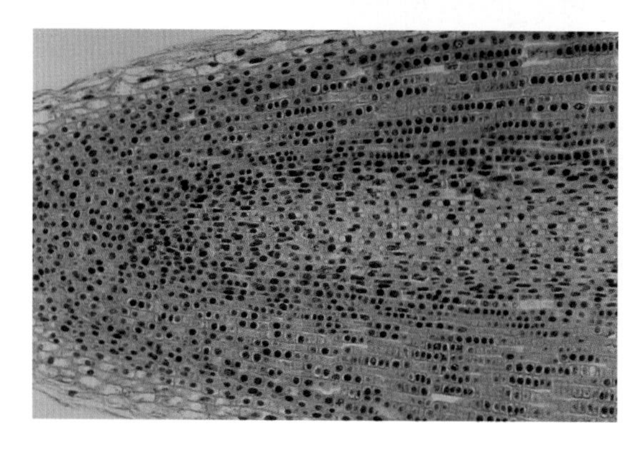

LEFT: **Root tip cells of vetch.**
RIGHT: **In plants, water and nutrients are drawn from the soil through the stem and along the leaves via xylem vessels, airtight tubes like these from an English oak tree.**

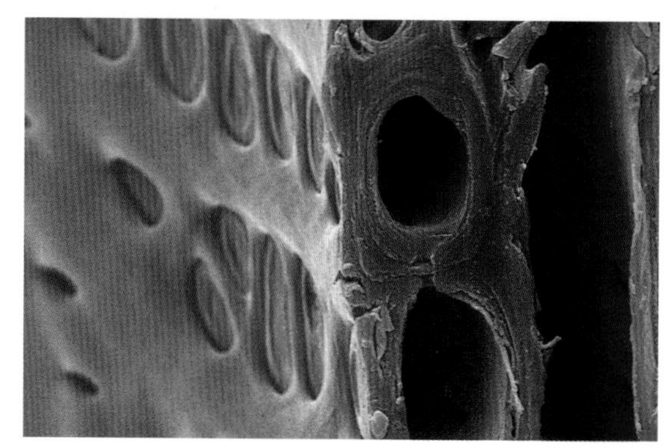

as humans are. Woese concluded that life consisted not of two but three domains, which he named Archaea, Bacteria and Eucarya.

This discovery immediately raised questions over which of the three domains are the most closely related to one another on the tree of life. Current research suggests that the split occurred over three billion years ago and that Archaea are more closely related to Eucarya than to Bacteria. Genetic sequencing confirms what the traditional fossil record has already suggested: that plants and animals are relatively recent branches in an already highly diversified tree, while micro-organisms represent most of the other branches, including all those at the bottom of the tree.

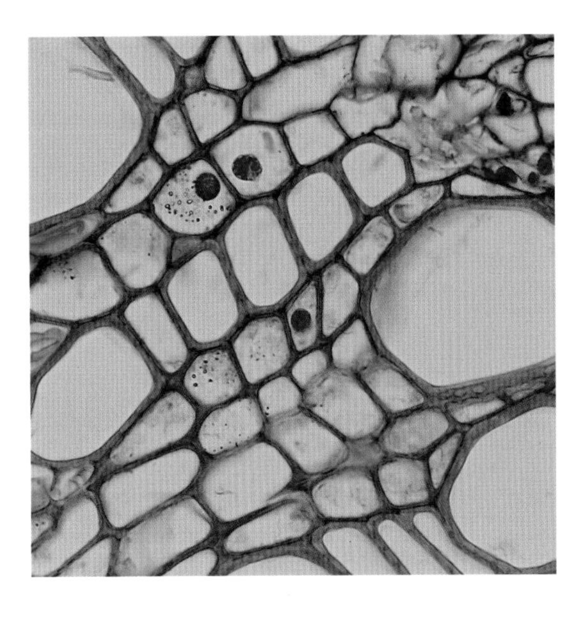

Root of broad bean showing thickened xylem cells.

CELL FACTS

- Most cells are impossible to see with the naked eye, the average width of a typical animal cell being only 0.02 mm, but there are exceptions. In giraffes, for example, some nerve cells can exceed lengths of over 4 m. Unfertilized ostrich eggs hold the record for the heaviest living single cells, but, remarkably, although they can weigh over 2 kg, they are among the smallest eggs, in relation to body size, laid by any bird.
- Cells are bound by a membrane that controls the passage of materials between the cell and its surrounding environment. Plant cells and bacteria are also enclosed by a tough, outer layer outside the membrane, called the cell wall.
- Some cells are able to move by waving minute hairs, called cilia or flagella (single: cilium, flagellum), that grow on their surface. Human sperm cells, for example, and a common, unicellular, freshwater organism, *Euglena viridis*, propel themselves by the whip-like action of a single flagellum.
- Plant and animal cells contain organelles — a range of complex structures that perform specialized functions, embedded in a jellylike substance called cytoplasm. Mitochondria, for example, release the energy from food that enables cells to operate and grow, whereas chloroplasts, specialized organelles only found in plant cells, carry out photosynthesis. DNA (deoxyribonucleic acid) controls cellular activity, and in all organisms except bacteria and archaea it is located within the chromosomes, contained within the nucleus.

The fresh water protozoan *Euglena* is normally found in puddles and often occurs in such large numbers that it colours the water green.

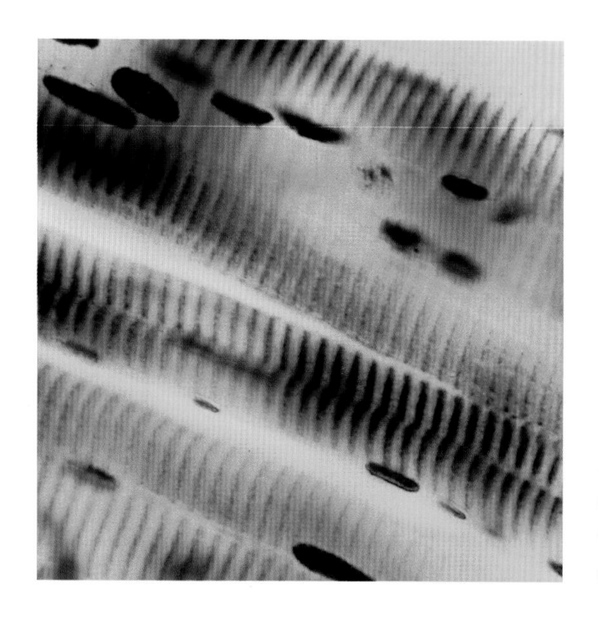

LEFT: **Striated muscle fibres.**

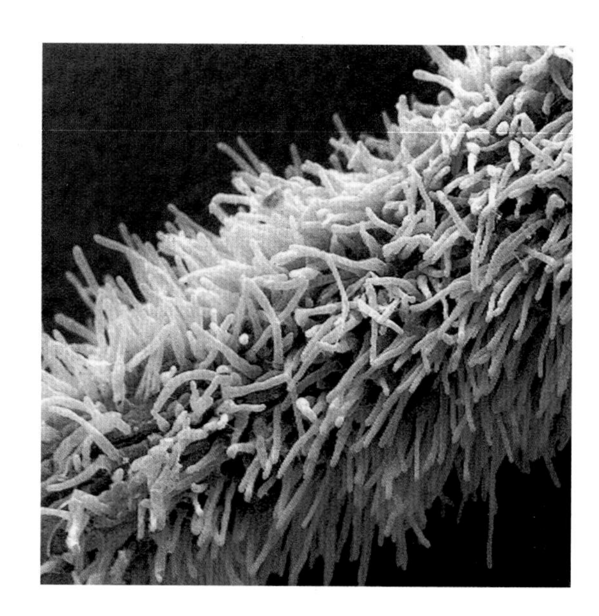

RIGHT: **Microscopic root hairs, like the ones shown here from oregano, increase the surface area through which plants absorb water.**

As the TEM (transmission electron microscope) images (below) show, plant and animal cells are structurally very different from bacterial and archaebacterial cells, although they are very similar in terms of their chemical processes. Bacterial cells are much simpler in their organization. They lack most of the organelles found in plant and animal cells, including a nucleus, instead retaining their DNA within the cytoplasm, in direct contact with the rest of the cell.

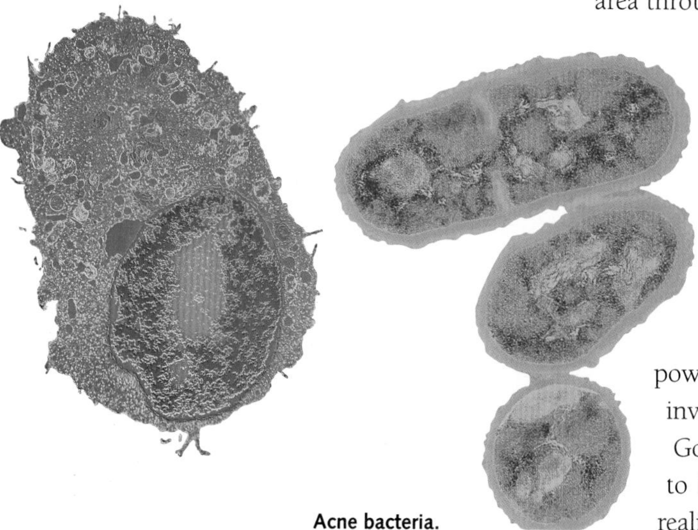

This colour TEM shows a single mammalian tissue culture cell. The large red shaded area is the nucleus and the blue area consists of cytoplasm and other organelles, including mitochondria, which are essential to the healthy functioning of the cell.

Acne bacteria.

Evolved for the Job

From the simplest molecules to the most complex organisms, there is evidence of the relationship between biological structure and function. Muscle cells, for example, consist almost entirely of bundles of specialized microfilaments that slide along one another to shorten the cells and produce a muscle contraction. Plant root tips are covered in minute root hairs, which maximize the surface area through which water and minerals may be absorbed, and giraffes have specialized valves in the blood vessels in their neck to prevent them from fainting when they hold their heads upright.

As with the example of the Galapagos finches in Chapter 2 (page 46), the correlation between structure and function can be a powerful tool in the hands of biologists investigating the natural world. When Gould identified the birds as belonging to 13 different species of finch, Darwin realized that the differences in the shapes

Susan Greenfield

PROFESSOR OF PHARMACOLOGY, OXFORD UNIVERSITY

Every one of us sleeps, feels pain, has felt the Sun on their face, has laughed. In short, every one of us has a brain. So the understanding of the human brain is one of the most fascinating areas of science, simply because it is all about ourselves.

The challenge of the human brain is to find out not only how drugs change moods, how you can make a thought translate into the mechanical contraction of muscle and hence a movement, or how you can interpret your surroundings, but also what makes you a unique individual.

The Natural History Museum is quite right in referring to human biology as an 'Exhibition of Ourselves'. In order to understand the most special and unique part of the body that is the brain, we must approach the problem from different disciplines and on different levels. It is no good just understanding its anatomy, or considering the brain as just a functional black box: instead, it is important to relate structure to function, to understand not just the anatomy but the physiology too. Moreover, we cannot just study the brain at one level; it is important to see how the gross brain regions relate to the actual tissue of which the brain is made and, indeed, how that tissue works from one moment to the next. The Museum achieves the complex and difficult link between understanding the close relationship between gross anatomy and the micro-circuitry and how they relate to the dynamic functioning of the brain.

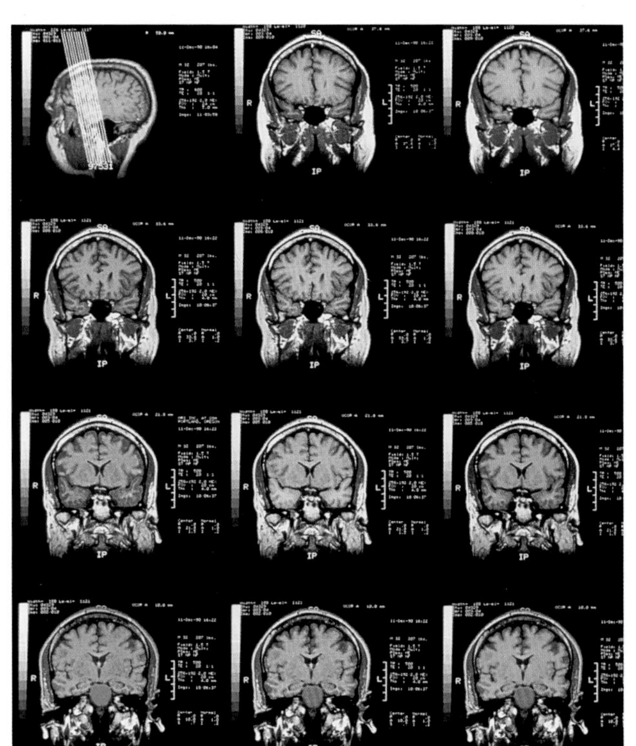

Coloured scans of a healthy human brain.

Paradoxically, the most intimate part of science, the understanding of ourselves, is also the most complex: the true 'final frontier'. The more we can share this vitally exciting question with young people and everyone else visiting the Museum, the more chance we will have of making progress in understanding who and what we really are.

DAYLIGHT ROBBERY AND THE TALE OF THE SWIMMING ANTS

Australian researchers have discovered why the world's only known species of swimming ants have decided to take the plunge. It might look beautiful, but the carnivorous pitcher plant *Nepenthes pervillei* is deadly to all insects lured into the fluid-filled cup. All insects, that is, except the *Colobopsis* ants that live in the long tendrils of the plant, which extend from the leaf–tips and end in the pitchers. Studies reveal that worker ants leave their home to swim through the plant's pool of digestive juices in search of large prey, such as crickets. As yet, no one knows how the ants climb back up the slippery surface of the cup to haul their stolen catch to the rim, so that other members of the colony can share in the feast. Remarkably, the plant benefits from this relationship because, if too many large insects accumulate, their decomposition fouls the digestive juices within the pitcher and prevents it from feeding. This symbiotic relationship between ant and plant provides a fascinating example of the way in which independent biological structures can evolve together to support the same function — in this case, feeding — in different organisms.

Nepenthes pervillei.

and sizes of the bills displayed by each species provided evidence of evolution in action. The evidence suggested that each species had developed a specialized bill, enabling it to exploit a specific source of food. The large ground finch, *Geospiza magnirostris,* for example, evolved a formidable beak, ideally suited to splitting tough seeds, in contrast to the long slender beak of the warbler finch, *Certhidea olivacea*, which feeds only on insects.

Certhidea olivacea.

Geospiza magnirostris.

INTERACTING WITH THE ENVIRONMENT

Living organisms react continuously with their environment, which includes other living organisms, as well as non-living factors like the climate, water availability and the chemical composition of the soil. These complex interactions occur at every level of the biological scale, and taken together form the basis of our understanding of natural communities.

OPEN BY DAY, CLOSED BY NIGHT

Plants have microscopic pores, called stomata (singular stoma), that control the flow of water, oxygen and carbon dioxide in and out of their leaves. Scanning electron micrographs show that each stoma consists of two specialized cells, called the guard cells, that become flaccid by night, sealing the pore, and swollen and taut by day, opening the pore. These responses are triggered by a combination of changes in the environment, and serve to help the plant function with maximum efficiency.

The opening and closing of the stomata reflect a delicate balancing act, allowing enough carbon dioxide for photosynthesis to flow into the plant but preventing too much water from leaving it. In plants that grow in temperate climates, opening the stomata by day allows precious water to evaporate, but also creates a powerful suction force that draws more water and vital nutrients up through the plant from the soil surrounding its roots.

This process, called transpiration, creates a stream that flows through the plant at rates of up to 75 cm per minute. At night, the dynamics change. Without sunlight to drive photosynthesis, there is no longer any benefit to be had from the absorption of carbon dioxide, and the stomata close, preventing any further escape of water from the plant. This pattern of behaviour may change, however, if the plant is exposed to significant changes in the environment, such as an exceptionally hot day. A rise in temperature will accelerate water evaporation from the leaves and increase the rate of transpiration and, if water loss begins to exceed water uptake, the stomata close to prevent the plant from wilting.

In desert climates, where temperatures can reach over 50°C and the average annual rainfall

Spiny cactus — *Ferocactus wislizenii.*

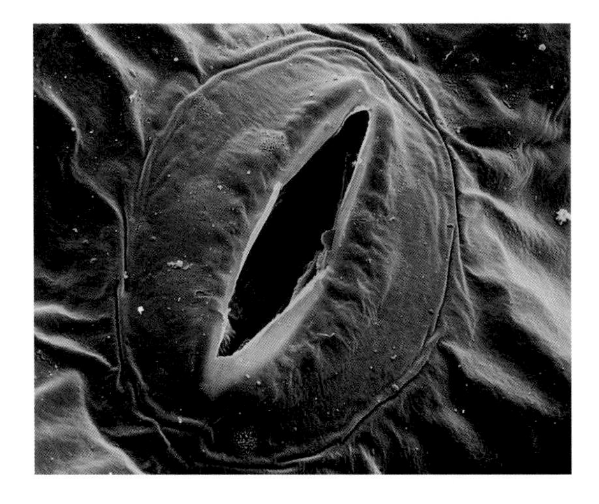

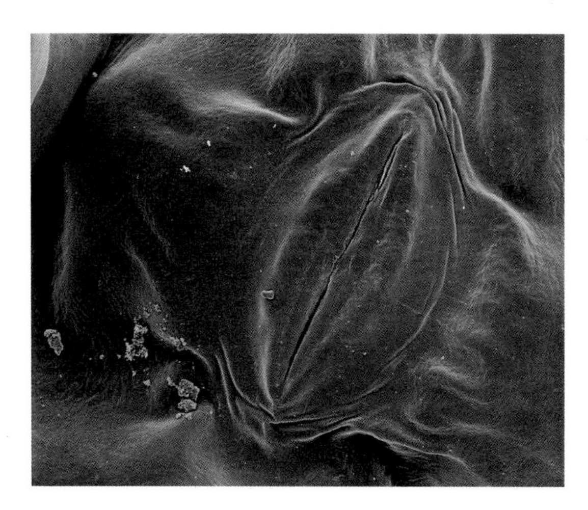

LEFT: **Open stoma.**
RIGHT: **Closed stoma.**

is less than 25 cm, water is at a premium, and the stomata in plants such as cacti behave in reverse. Specially adapted to survive in their inhospitable environment, species like the American *Ferocactus wislizenii* have leaves reduced to mere spines and carry out photosynthesis in their modified stem. To prevent excessive loss of water during the heat of the day, they open their stomata only at night, storing carbon dioxide in their stem ready for photosynthesis the next day.

A wood of sessile oak.

Ecology and Ecosystems

An ecosystem may include hundreds or thousands of different species linked together by a variety of food chains. It can range in size from a single tree to a forest or a coral reef, culminating with the complete biosphere, at the grandest level of all.

Ecosystems are based on two key processes, chemical recycling and energy flow. In most cases, energy enters an ecosystem in the form of sunlight. Along with water and carbon dioxide, this is converted by photosynthetic organisms, such as plants, algae and some bacteria, into organic sugars, forming a chemical store of energy. When animals eat plants and other photosynthesizers, or other animals, the organic compounds are digested, releasing energy back into the atmosphere as heat. Chemical recycling occurs when, for example, the minerals absorbed by plants are returned to the soil by the action of microorganisms that can break down material such as leaf mould, animal remains and faeces.

ECOLOGY — PAST AND PRESENT

Ecology — understanding the relationships between different species and between them and their environment — is important because it helps scientists to understand the implication of change on species living in the wild. These changes may result from human activities, including deforestation, urban development and industrialization. Alternatively, they may be natural, resulting from sudden catastrophes, such as a volcanic eruption or meteorite impact, or from longer-term climate change. Moreover, immigration of species into a new area may cause displacement of existing fauna or flora; this may happen naturally or be human-induced. An understanding of the effect of these activities is the first step towards finding solutions to control or, at the very least, minimize, the impact on the natural world.

Although most ecologists study the present-day interactions of living organisms with their environments, they can also learn much about the impact of environmental change over hundreds of millions of years on ecosystems that flourished in ancient geological time.

Palaeontologists Jerry Hooker, from The Natural History Museum, and Margaret Collinson, from Royal Holloway College, University of London, for example, are investigating the interactions between plants and plant-eating mammals, in the thirty million years that followed the mass extinction of the dinosaurs at the boundary between the end of the Cretaceous period and the beginning of the Teriary period, 65 million years ago. Their research focuses on the Euro-American region, and is based on a wide range of fossil evidence, including pollen grains, fruits, seeds, leaves and large tree stumps from plants, and isolated teeth, jaws, limb bones and rare skeletons from plant-eating mammals.

Using this fossil evidence, the two scientists have been able to reconstruct the changes in the ancient vegetation and climate, together with the corresponding changes in the diet and feeding behaviour of mammals. So far, the research shows that after the mass extinction of the dinosaurs, mammals diversified to exploit the fruits, seeds and leaves of plants that had previously provided a primary source of food for many species of plant-eating dinosaurs.

Before the Cretaceous-Tertiary boundary, mammals were small and fed mainly on insects, but after the extinction of the dinosaurs, larger, mainly ground-dwelling mammals, which would have subsisted largely on fallen fruits, began to evolve. Modern-day examples of this ecological type are peccaries and chevrotains.

Such ecological types were common until late in the Eocene period, about 40 million years ago, but are rare and essentially restricted to the tropics today, where the constantly warm climate means a year-round supply of fruit. Their almost complete disappearance seems likely to be due partly to the cooling climate, which meant that fruit became scarce at higher latitudes. It was also partly due to the evolution of new fruit types: those with soft flesh, and those with a hard outer casing (nuts), which evolved in response to the benefits to the plants of seed dispersal by primates and rodents respectively.

Collared peccary from southern Texas.

Consequently, fruits were either foraged at source in the trees or their outer layer was too hard to be opened by most animals, apart from rodents. Such co-evolution between plants and mammals appears to have become prominent by the Eocene.

This research also shows that, despite the successful proliferation of flowering plants during the Cretaceous period, there was a surprising gap of at least a million years, after

Chimpanzee eating berries.

the dinosaurs perished, before mammals began to exploit leaves as a source of food. There may have been several reasons, but Jerry Hooker and Margaret Collinson believe that the delay may have been partly due to the favourable climate during the Palaeocene epoch, which followed the Cretaceous period. This encouraged a transformation of the open canopy woodland of the late Cretaceous into rainforest. Such a change would have led to an abundance of fruit and dense vegetation, which may have discouraged the evolution of large mammals. Digestion of leaves involves breakdown of cellulose by bacteria; this requires a long digestion time, easily achieved only by an increase in size, providing a longer gut.

This theory is supported by the fact that, despite the appearance of the first mammalian herbivores in the Middle Palaeocene, about 62 million years ago, there were relatively few species until a dramatic expansion 25 million years later, during the late Eocene. By this time, climates were cooling significantly, and the green parts of plants became a more reliable food source than fruit in the increasingly seasonal climate and more open vegetation.

Amoeba — asexual reproduction.

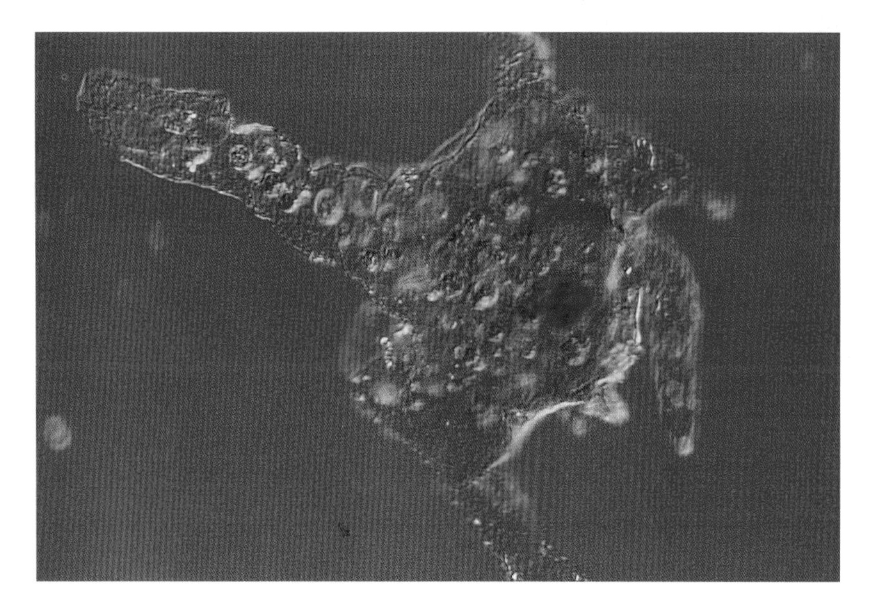

Like Begets Like

What else could bacteria, a sunflower, a human and a whale possibly have in common? We have seen that despite the enormous diversity of living organisms, they all share the same basic characteristics. They are all made of cells, are highly organized, consist of complex structures evolved to perform specific tasks, and interact with the environment. But there is something else, at an even more fundamental level, that all organisms share in common. Their cells contain molecules of DNA, the chemical code that controls all the metabolic processes necessary to sustain life, and the mechanism by which biological information is passed from one generation to the next.

DNA consists of two spirals, or chains, made up of four basic building blocks, smaller components called nucleotides. The chains are linked together by hydrogen bonds that form between the nucleotides on opposite chains, and give DNA its characteristic double helix shape. The precise sequence of nucleotides corresponds to specific instructions called genes, like a library of commands which the cell 'reads'.

In eukaryotic cells, genes exist within long threadlike structures within the nucleus, called chromosomes, which replicate, to make a precise copy of each gene, every time a cell divides. Cell division occurs within our bodies all the time. It increases cell numbers, and is the process by which we grow and our tissues are repaired or replaced. It is also the mechanism by which some organisms reproduce.

Reproduction guarantees the continuation of life, and it is one of the key processes that distinguish the living from the non-living world. There are two methods of reproduction, asexual and sexual.

Asexual reproduction involves a single parent, either splitting in two by a process

GENE TECHNOLOGY

Gene technology is also referred to as genetic engineering or biotechnology, and describes the growing field of research that explores the transfer of genes from one organism to another. This process has two important applications. It can be used to produce large amounts of proteins, to be used, for example, in medicine or for making food — such as insulin to treat diabetes and rennin to make cheese. It can also be used to breed organisms with altered characteristics such as disease resistance in plants. The application of gene technology has huge commercial potential, and has led to the rapid development of the biotechnology industry in recent years, but it has also raised considerable concern over the safety and ethical implications of genetic engineering.

In Europe, recent scares about the safety of eating genetically modified (GM) food has led to mass demonstrations. In countries such as Germany and Britain, all the major supermarket chains label products that contain genetically modified ingredients. Organic farmers and environmentalists are also concerned about the implications of growing GM crops alongside natural plants. British botanists, for example, recently found that the pollen produced from fields of GM oilseed rape remains fertile over much greater distances than was previously expected — approximately twice the required distance of the 'buffer zones' that must exist between transgenic and natural plants, according to the rules that currently regulate the farming of GM crops in Britain. Environmentalists are concerned that the genes that introduce resistance to herbicide in the oilseed rape might spread to nearby weeds, and organic farmers fear that the airborne pollen might affect the status of their farms.

Ethical considerations focus on what is and what is not acceptable in genetic engineering: the level of suffering that may be caused to transgenic animals, for example, and the moral implications of cloning organisms like Dolly the sheep. Should transgenic animals be cloned to manufacture products that may help combat human diseases such as cystic fibrosis? Should scientists apply cloning techniques to humans? Should we demand more stringent tests to assess the safety of cultivating and eating GM food? As our knowledge of genetic engineering advances, some extremely important and difficult decisions must be made.

Should our food be genetically modified?

WEIRD, WILD AND WONDERFUL ANIMAL SEX

Many marine invertebrates and some species of freshwater and terrestrial vertebrates reproduce externally. The parents discharge sperm and egg cells, and fusion occurs in the surrounding water. Many species attract a partner through elaborate courtship rituals. The male three-spined stickleback, for example, performs a series of underwater dances and changes colour, while male dragonflies court females by altering their flight pattern. Mating in dragonflies requires considerable gymnastic ability on the part of the female. It begins when the male grasps the female's head with his abdominal claspers and she loops her abdomen over to position their sexual organs together.

Elaborate courtship displays may be important to animals that use natural Sunlight to see one another, but how does a deep ocean organism that lives in the pitch dark find and recognize a potential mate, let alone signal to it its desire to mate? About 80% of deep-sea organisms are capable of emitting light, either by themselves or with the help of symbiotic bacteria, and biologists have discovered that fish living in the deep ocean rely on this bioluminescence to signal to one another in the dark.

One group, the family Stannidae, known as dragon fishes because of their powerful jaws and long sharp teeth, emit light from organs on the head and body that act as lamps. Most bioluminescent organisms emit blue-green light, but the fish in the subfamily Malacosteinae, commonly known as loose-jaws, are the only ocean animals known to emit and detect red light, which allows them to hunt prey and signal to prospective mates without being seen. Some deep-sea angler fish have found a simple solution to the problem of locating a mate in the deep ocean: once they find a suitable partner, they never let go. The much smaller male takes up permanent residence, attaching himself to the female with his jaws. His blood system links up with hers, his heart withers, and he becomes totally dependent on her, his function only to produce sperm.

Japanese crane courting ritual.

Land-based couplings are constrained by the lack of water, and most animals overcome the problem by reproducing internally. Internal fertilization occurs when the male injects sperm directly into the body of the female. Some animals, however, still rely on external fertilization, and have found other methods for exchanging their sex cells. Grey slugs, *Limax maximus*, for example, are hermaphrodites (with both male and female reproductive organs in the same individual) that twine together at the end of a rope of slime so that they can exchange sperm at the end of their extended penises. After copulation, the slugs separate, before eventually both laying fertilized eggs.

Sea horse parent and young.

called fission or forming buds that break away to establish themselves as new individuals. Asexual reproduction is fast and effective. It doesn't involve searching for a suitable mate, or the lengthy courtship rituals that tend to accompany successful sexual liaisons, but it does have disadvantages, too. In most cases, the resultant offspring are genetically identical to their parent, and therefore share the same strengths and weaknesses.

Sexual reproduction occurs when a sperm, a specialized sex cell containing genes donated by the male parent, fuses with an egg cell, a specialized sex cell containing genes of the female parent. Fusion produces genetically varied offspring that inherit unique combinations of genes from both parents. For a long time, biologists believed that, because this variation produces some individuals within populations that are better adapted to changes in the environment, sexual reproduction offers a significant advantage over asexual methods.

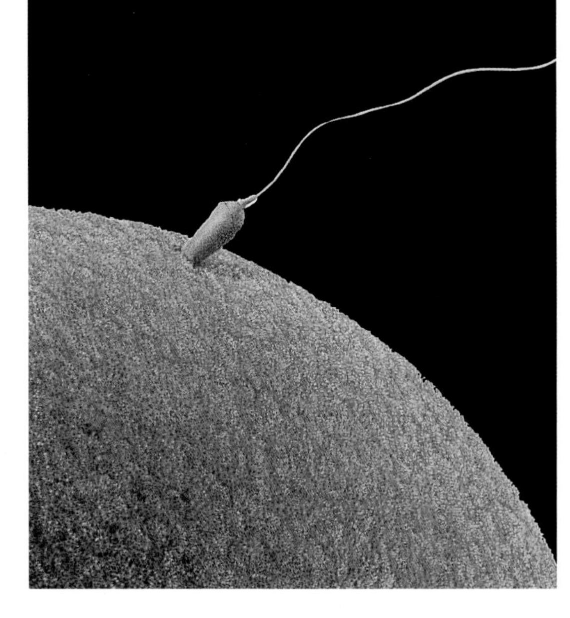

The relatively tiny human sperm cell in the process of fertilising an egg.

The theory suggests that the individuals with the advantageous genetic variation will be able to survive adverse changes in the environment, so although those that lack it will perish, ultimately the population will continue, through the offspring of the 'fittest' who have survived. This assumption has been hotly debated over recent years, and it seems that the relative merits of each method may not be as obvious as was once assumed.

Diverse Yet Not So Very Different

Despite centuries of exploration and collection, producing an ever-increasing record of the overwhelming diversity of life, many biologists believe that we are still familiar with only a tiny fraction of the living organisms that populate our planet. The Earth is vast, and there are still many relatively inaccessible regions, such as the ocean depths and the Earth's crust, that we are only just beginning to understand. Every year, biologists identify thousands of species new to science, adding to the extensive list of over two million that have already been named.

The species that occupy the Earth today are as diverse as they are numerous, the product of over 3800 million years of evolution, operating in response to the enormous physical diversity of the Earth's surface. They come in all shapes and sizes, and exploit nearly every conceivable habitat, no matter how small, and no matter how extreme the conditions. Until twenty years ago, for example, most biologists thought that the maximum sustainable temperature for living organisms was 50°C, but scientists have since discovered species of thermophilic (heat-loving) and hyperthermophilic (superheat-loving) organisms, including bacteria and Archaea, that can happily survive much higher temperatures. The upper temperature limit of life on Earth is the subject of current research. Evidence suggests that life exists in excess of 140°C.

DIVERSITY IN PELARGONIUMS

You may expect such obviously diverse organisms as an elephant, a spider and a daisy to look and behave very differently, but there can be considerable variation among species of the same family, too. Mary Gibby, a botanist who specializes in studying pelargoniums, explains:

'Pelargoniums are well-known to gardeners in the form of the scarlet geranium, trailing ivy-leaved geraniums, scented geraniums and regal pelargoniums. But as well as these cultivated forms (cultivars), there are many more varied forms in the wild. In fact, there are well over 250 species of *Pelargonium*, and most of them are found in the south-western part of South Africa. Here, we find wild species, such as *Pelargonium zonale* and *P. peltatum*, that look fairly similar to our garden cultivars but, in addition, we find succulents (*P. carnosum* and *P. laxum*), tiny, weedy species (*P. nanum* and *P. minimum*), tuberous-rooted species (*P. pilosellifolium*, *P. rapaceum* and *P. triste*) and the strange, semi-succulent, scrambling *P. tetragonum*. Most of these plants grow in a region where there is winter rainfall followed by a very hot, dry summer, and it appears that many pelargoniums have become adapted in different ways to cope with the same problem of seasonal drought. Such plants may be short-lived annuals that produce a lot of seed to survive the dry season, they may have succulent stems to retain moisture, or their aerial parts may wither, leaving them to survive the summer beneath the ground as tubers.

Although some species of *Pelargonium* are found further north in the mountains of East Africa, the Arabian Peninsula, Turkey and Syria, most live in the southern hemisphere, in southern Africa, Madagascar, Australia, New Zealand, and the Atlantic Ocean islands of Tristan da Cunha and St Helena. Of these species, probably 80% are concentrated in the winter rainfall area in the south and west of South Africa and Namibia.

This 'centre of diversity' for *Pelargonium* was thought at one time to also represent the 'centre of origin' – a place where *Pelargonium* had its origin. But more recent work, using DNA sequence data to construct a phylogeny (evolutionary history) of *Pelargonium*, indicates that the great number of species that occurs in the winter rainfall area may have evolved relatively recently, in response to the drying out of the climate since the late Tertiary period.'

ABOVE: ***Pelargonium trifoliatum.***

ABOVE: ***Pelargonium multibracteatum.***

ABOVE: ***Pelargonium bowkeri.***

Chris Packham

NATURALIST AND BROADCASTER

I distinctly remember my first visits to The Natural History Museum; they always seemed to be on rainy days, when its dark yet auspicious halls provided a cavernous and fascinating escape from the inclement weather outside. As with so many other people's recollections of discovering the Museum, it is the sauropod skeleton that is lodged most prominently in my mind; it is only when you see the skeletons of great dinosaurs 'in the bone' that you truly appreciate their awesome scale. Scale, it seems to me, is always missing, or difficult to make out, from books, television or CD-ROMs, and as a small child gazing up at this ancient behemoth, I was immediately struck by the immensity of these extinct animals.

A mass of monarch butterflies, Mexico.

On one occasion, years later, we were fortunate to be making a television programme about dinosaurs, and were filming late in the Museum, after the public had left. It was dark, and only a few lights were on. A good friend of mine, the herpetologist Mark O'Shea, had brought with him his pet dwarf caiman for us to use when discussing the extinction of the dinosaurs and those groups of animals which survived the catastrophe. During a break in filming, I was sitting in the Central Hall, re-writing a few pieces of script, when I looked up to see Mark walking beneath the great *Diplodocus*. Following him like a well-trained dog, only a few feet behind his heels, was his caiman, and to see the juxta-position of man, giant extinct reptile, and beautiful living reptile has left a lasting image in my mind.

The natural world continues to amaze me. I suppose of the animals, hyperparasites (parasites of parasites) are my favourites, since the complexities of their life histories are totally intriguing, and I wonder at the mechanisms of their evolution — what tiny twists occurred in their genes that enabled them to develop such bizarre combinations of behaviour? But I am also deeply affected by the spectacle of any large aggregation of animals, since today these sadly appear only as relics of a bygone and wondrous age. The sight of five million flamingos in Africa's Rift Valley and 50 million monarch butterflies in the mountains of Mexico have both left me quite speechless.

Precisely because such amazing sights are becoming ever rarer, one issue always in the news today that concerns me greatly is the reluctance of many people to begin to under-stand biologists' exploration of genetic science. In many instances, an unreasoning fear of 'genetically manipulated' organisms may turn out to be a catastrophe for nature — and ourselves. Evolution is a process of natural genetic manipulation, and it's slow, it's careful and it never makes lasting mistakes. I am sure that if people would only take an unbiased look at the evidence, then they would realize that some of the new techniques of genetic manipulation, coupled with the necessary caution, could prove of immense benefit to the countless species of animals and plants with which we share this planet.

A termite mound in Nigeria.

STILL THE SAME SPECIES: MACROTERMES

In most species of plants and animals, the range of variation between individuals is very small, but for some the differences can be so striking that it is difficult to believe they are related. Termites, for example, have evolved complex caste systems based on a range of individual types within a single species (polymorphism), with each type adapted to perform a specific task. An example of this is in the African species *Macrotermes*. The different types within a species are shown on the right. The large egg-laying queen and the smaller king are in the middle; these are surrounded by white larvae, and a variety of sterile types, including blind wingless workers with round heads and small jaws, and soldiers with oval heads and enlarged jaws specifically adapted for defence.

They May Look Similar but are They Related?

Plants and animals living in similar environments face the same problems. Not surprisingly, adaptations to overcome these problems often produce strikingly similar features in form as well as function, even in completely unrelated species. This is known as convergent evolution.

Heaths and moorlands are open habitats, with very acidic soils that are low in minerals and badly aerated. Plants growing here must contend with exposure to strong winds, rapid water loss and low nutrient levels. The growth form which most successfully meets these requirements is a low, tough-stemmed shrub with evergreen leaves that are small and narrow, often with inrolled margins — all features that reduce surface area and help to curb water loss from the leaves. Stomata (the pores through which atmospheric gases are exchanged) are sunk into pits or lie in grooves to further prevent evaporation. Hairs on leaves and stems help to trap a layer of moist air next to the plant's surface.

This overall form is referred to as ericoid, because it is epitomized by species of *Erica,* the heaths and bell-heathers. *Calluna vulgaris*

Polymorphism in a termite species.

(heather) and species of *Vaccinium* (bilberries) are other ericoid members of the heath family. Another dominant group on heath and moorland consists of *Empetrum* species (crowberries). Although belonging to a completely different family, they too exhibit precisely the same ericoid features. In fact, so similar are many ericoid species that it can be difficult to distinguish them without close study. The convergence goes beyond overall appearance. Members of the Ericaceae lack root hairs, their function being taken over by mycorrhiza — symbiotic fungi that are bound up in an intimate association with the plants, to their mutual benefit. The efficiency of the mycorrhiza and their ability to perform in waterlogged, badly aerated soils is a major advantage. Members of the Empetraceae face identical problems, and have arrived at an identical solution. They, too, have mycorrhizal fungi associated with their roots.

DIFFERENCE DURING DEVELOPMENT

All plants and animals change size and shape during different phases of their life cycle. The South American paradoxical frog, *Pseudis paradoxa*, provides a remarkable exception to the general rule that organisms tend to increase in size with age. It earns its name from the fact that the tadpoles can grow to nearly four times the length of the adult frogs. The mature tadpoles gain most of their 250 mm length from their tails, which are absorbed when they transform into much smaller frogs that reach lengths of only 70 mm.

By contrast, the example of *Carica papaya* (below), the plant that produces the edible fruits known as papayas or paw paws, shows the progression of development from juvenile to

This series of images (left to right) shows the development of *Carica papaya* from seedling to adult.

HOW MANY SPECIES ARE THERE IN THE WORLD?

Counting species may not be as straightforward as we might assume. When we want to measure something, it is important that we agree on the units of measurement. For example, a class of schoolchildren asked to measure the width of their classroom with rulers, some in centimetres and some in inches, will arrive at two different results. The numbers of each unit they measure hold some information about size, but without an understanding of the units they are of very limited value. The children's measurements become meaningful only when we know which scale was used, and how the different scales relate to one another. In modern biology we talk about the diversity of an environment, and we tend to use the number of species as a measure of that diversity, but how reliable are the units we are counting?

WHAT IS A SPECIES?

The most familiar definition of species is generally known as the 'biological species concept'. It states that two individuals belong to the same species if they can interbreed to produce fertile offspring. However, this is only partially true. Under normal circumstances, individuals of the same sex won't mate to produce fertile offspring, even though they belong to the same species. This definition also implies that organisms reproduce sexually, creating offspring that share genetic material from both parents. While this is true for many of the larger, familiar plants and animals, the definition breaks down when we turn to organisms that are asexual in their reproduction, including bacteria, some fungi, and even a few animals and plants — the banana plant, for example.

Banana plant.

For organisms that are asexual the concept of species is far less rigid, and the only generally accepted definition is: a species is a group of organisms that we find it convenient to name. Because this definition is open to interpretation, the term 'species' cannot be taken to convey an equivalent unit of measurement. Comparing estimates of the total numbers of asexual 'species' is even more meaningless than comparing measurements made in inches and centimetres, since it is without the benefit of a conversion factor.

For example, there are a number of variants of *Escherichia coli*, the bacterium that lives in the digestive tract of mammals. Bacteriologists recognize different variants, or strains, of *E.coli*, in much the same way that gardeners identify different varieties of roses. It would be reasonable to expect that there would be a high degree of similarity among these strains, but comparisons show that the genetic similarity within *E.coli* is only about 70%. This is far less than the genetic overlap recorded between entirely separate species: the genetic overlap between chimpanzees and humans, for example, is more than 99%.

This is the crux of the problem: the simple measure of similarity is not enough. It is not merely the percentage of difference, but where those differences fall. This, in turn, depends on which genes the organism carries and, as we do not yet have a sufficient understanding of genome structure, we do not know whether it will ever be possible to build a usable definition of species in this way.

BIODIVERSITY

We often hear or read about biodiversity, mostly in relation to issues concerning extinction of species and loss of habitats, but it is a complex phenomenon, difficult to define and measure in any simple way. One definition of biodiversity includes the species that are active within a habitat at any given time, those that are not presently active because

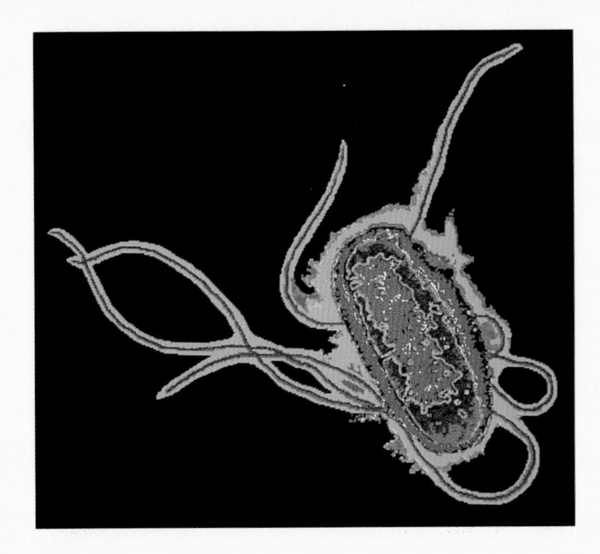

E.coli.

they are dormant or migratory, and the habitat system as a whole, including the interactions between the various species, food webs, nutrient cycles and neighbouring habitats.

The number of species active in any habitat varies, depending on the environmental conditions now and in the recent past. In winter, for instance, many organisms are dormant and may not be detectable, but will appear when the temperature or day-length increases, or until some other environmental factor, such as a rare shower of rain in a desert, awakens them. Plant seeds can survive for years before the right environmental conditions trigger germination, and bears and other mammals hibernate to conserve energy. Change in the environment may also attract organisms from other habitats. Cultivating a garden around a brand-new house, for example, will attract many species of insects, including butterflies and bees attracted to the nectar produced by flowering plants.

But how can we begin to measure biodiversity unless we have a clear definition of species? As we saw earlier, the difference between species ultimately boils down to differences between their genes. Entomologists estimate that there are between 3 and 30 million species of insects, all belonging to a single class of arthropod, constructed from a rather limited set of genes on a single body-plan. But the total genetic similarity among these insects is of the same order as the similarity found within *E.coli*, just a single species of bacterium.

CONSERVATION

So what does this tell us about conservation? When setting conservation targets, what exactly are we trying to conserve? Ecosystems are under threat all over the world from mankind's changes in patterns of use and, in general, it seems that the more diverse habitats are more resilient to change. We don't as yet fully understand the components of this diversity, which is why habitat preservation is so important, but ultimately, it may well prove that genetic diversity is the key. And if this is the case we need to formulate a way of judging whether it is better to lose 1000 species that are very similar to species elsewhere, or a single species that is genetically very different from anything else known. At present, science has a very poor understanding of the true extent of genetic diversity because we have a biased sample from the tree of life. Rainforests are considered diversity hot spots, but like any terrestrial habitat they contain only 19 eukaryotic phyla. The oceans, on the other hand, contain representatives of all 36 animal phyla, but are far less well known.

SPECIES COUNTS

As we cannot define the unit of measurement, it is impossible to estimate the total number of species in the world and different scientists make very different estimates. But this question may be less important than being able to make useful comparisons. Asking how many known species of birds there are in the world, for example, and comparing this to the number represented in a particular habitat gives us usable information, both about the diversity of birds and about the habitats they occupy. Such counts can be used as indicators of the environment's health and stability, and are an essential tool of conservation management.

How many species are there in the world? This may not even be a sensible question to ask, but trying to measure and preserve diversity will continue to be important.

MINERAL DIVERSITY

A staggering 85% of the Earth is made up of crystals, mostly in the form of minerals. They form all the rocks within the Earth and on its surface, and occur in a variety of structures found in living organisms, including bone, teeth, feathers and shells. Naturally occurring minerals and man-made crystals are all about you. They 'fur' your kettle, rust your car, and are added to the paper that wraps your gifts. Minerals are cut and polished to make gemstones for jewellery, crushed to produce the abrasive grit that covers the tip of a dentist's drill, and manufactured to make the silicon chips in digital watches and personal computers.

Mineralogists have already distinguished 3700 species of minerals, and discover about 30 to 40 new ones each year. Defined as inorganic, naturally occurring, crystalline solids, each mineral has its own unique chemical composition and structure based on a regular pattern of atoms, This gives it distinctive characteristics when tested in the laboratory and examined under high magnification.

The honey coloured crystals of scalenohedral ('dog-tooth') calcite.

Oxygen and silicon are the most common elements in the Earth's crust, and they bind together to form hard crystals called silicates. Silicates combine with other common elements, such as aluminium, calcium, iron, magnesium , sodium and potassium, to make up most of the rocks found on Earth. They include feldspar, mica, quartz, and sandstone. Important non-silicate minerals include calcite (the basis of limestone and marble), diamond (a form of crystalline carbon and the hardest substance known), and ice.

Quartz under the microscope.

Minerals, like living organisms, are extremely diverse. No crystal is perfectly pure, and these impurities present limitless possibilities for variation in their colour and appearance. Impurities might arise from the addition of a rogue atom, or from bubbles of gas or liquid, trapped within the crystal during its formation. Very subtle changes, such as the substitution of tiny proportions of the silicon with iron in quartz, can dramatically alter the appearance of a mineral to the naked eye. This slight alteration in atomic structure changes the amount of light absorbed by the crystal structure, and turns the transparent quartz a stunning purple. It looks like a completely different substance, and is popularly known as amethyst, but essentially its chemical properties remain the same, and it is still a piece of quartz. Similarly, milky quartz looks opaque to the naked eye, and if you examine it under a powerful light microscope you can see the cause of its opacity — millions of tiny fluid-containing gas bubbles or, in some cases, millions of crystal fibres of other minerals encapsulated in the quartz.

adult plant. The first 10-20 leaves are rather egg-shaped, with a smooth margin. Later leaves are progressively more lobed: at first, the lobes are scarcely more than a pair of large teeth along the margin, but these gradually become more distinct and the number of lobes increases. Adult leaves have up to 9 lobes, each of which is itself lobed. There is also a great increase in size from the first small seedling leaves, only 1–2 cm long, to the adult leaves, which can be up to 40 cm long.

The lack of resemblance between juveniles and adults in many plants creates an interesting and often frustrating challenge for botanists trying to identify juvenile plants in the field. How can they correctly identify individuals at these different stages of development when they look so little like the adults described in floras? This question was put to Nancy Garwood, a botanist who specializes in studying Central American plants.

'Years of experience comparing hundreds of species is an important element of the detective work involved, but the only way to be absolutely sure is to harvest seeds from the adult plant and observe and record the pattern of development as it grows. I have done this for more than 750 species in Panama, and am publishing a seedling identification guide so that others can name these stages more easily.'

COCCOLITHOPHORIDS

In the open ocean the primary producers, the base of the food chain, are single-celled plant plankton, such as diatoms, dinoflagellates and coccolithophorids. These phytoplankton thus have a role in marine ecology analogous to that of plants on land. So, understanding their biology is central to the study of oceanic ecosystems and to modelling the oceans' role in the carbon cycle.

However, the microscopic nature of these organisms makes direct observations difficult. Single cells can rarely be watched for more than an hour or so, and often do not show any very informative behaviour. In particular, this means that it is often difficult to work out the lifecycles of these organisms. With land plants, botanists can usually record the development of a plant as it grows from seed to adult. With phytoplankton, asexual reproduction by simple binary division of cells, producing two similar daughter cells, is commonly observable, but complete sexual life cycles are likely to involve an alternation of generations — which may never occur in the rather artificial conditions of a microscope preparation.

Coccolithophorids form an interesting example of this alternation of generations. They are one of the main phytoplankton groups, characterized by possessing an exoskeleton (external skeleton) of intricate calcareous plates, called coccoliths. There are two basic coccolith types: holococcoliths, formed from numerous minute simple crystals, and heterococcoliths, formed from radial cycles of larger, morphologically complex, crystal units. Usually, coccolithophorids form coccospheres (remains of the coccoliths) that have only one of these coccolith types. However, in the particular case of the species *Coccolithus pelagicus*, it has long been known that the two types are produced during alternate phases of the life cycle. This was first observed in laboratory cultures; subsequently, in oceanic samples, occasional 'combination coccospheres' of *C. pelagicus* have been observed that have both coccolith types, evidently recording the life-history transition. While this example found its way into the textbooks, it was by no means clear whether this was a typical pattern for the group as a whole, or an unusual evolutionary relic.

Recently, there has been renewed interest in coccolithophorids as key organisms in global change studies, particularly since their calcification (in which they become hardened by impregnation with calcium salts) plays a key role in the marine carbon cycle. For example, an EC-funded project, CODENET (the Coccolithophorid Evolutionary Biodiversity and Ecology Network), unites eight leading European laboratories (led by The Natural History Museum) to investigate outstanding problems related to the phylogeny (evolutionary history), microevolution (change due to mutation over a relatively short time) and ecology of these important planktonic algae.

In 1998, Lluisa Cros and Jose Fortuno, from the Barcelona Institute of Marine Sciences, were working on a detailed investigation of coccolithophorid biogeography in the Western Mediterranean, as part of the CODENET project. They were surprised to find examples of combination coccospheres, containing heterococcoliths and holococcoliths, in species never previously known to have complex life cycles. A painstaking search of their samples revealed several more examples.

Working with other CODENET scientists, Cros and Fortuno have now able to show that these examples come from virtually all known families of coccolithophorids, and so predict that two-phase life-cycles are a common feature of this group. The next challenges are to test whether genetic recombination occurs during this type of life cycle, and to interpret the ecological significance of the two phases. This is crucial. If scientists do not understand the basic biology of organisms, they cannot hope to interpret their role in the ecosystem meaningfully.

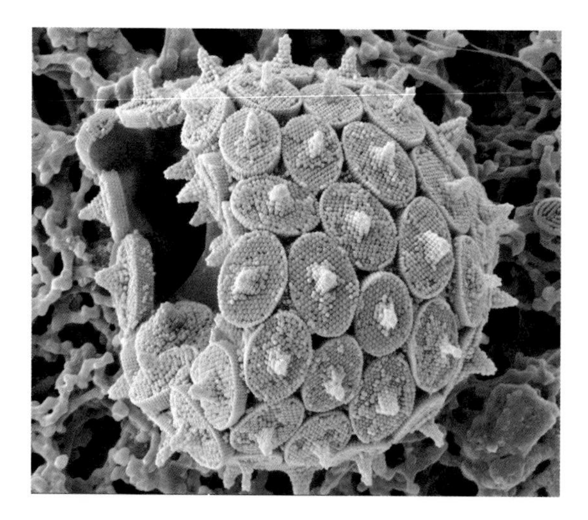

Helicosphaera carteri — holococcolith cells.

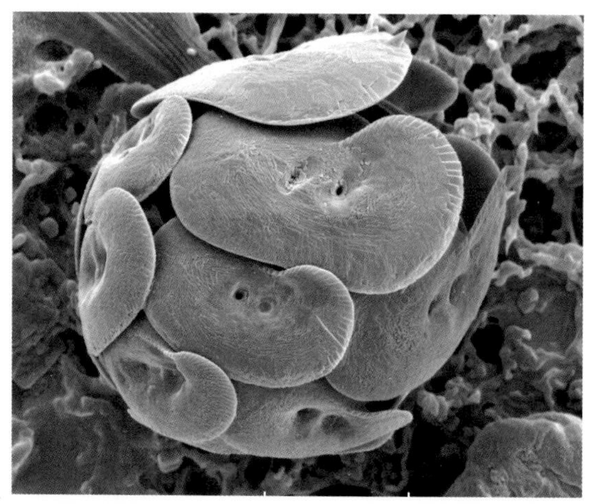

Helicosphaera carteri — heteroccolith cells.

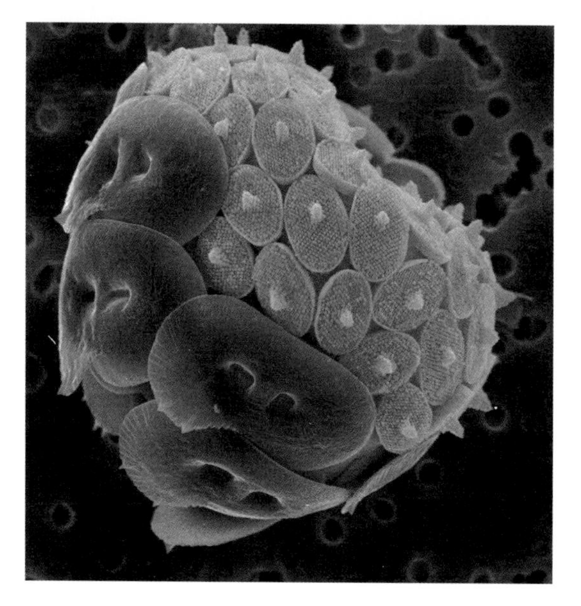

Helicosphaera carteri — combination cells.

A Single Common Ancestor

Darwin was prophetic when he first proposed that all living organisms had evolved from a single common ancestor in a succession of new species diverging like the branches on a tree spreading outwards from a single trunk, but today many biologists are convinced that this assumption is correct. As the examples in this chapter have shown, living organisms, despite their huge diversity, share too many complex features in common for all of these features to have arisen over and over again on many separate occasions.

This doesn't necessarily mean that all life has descended from the first living thing. In Darwin's tree of life, the branches at the top of the tree represent present-day species, while the trunk represents the first living organism. The common ancestor sits at the last fork, from which all the branches corresponding to present-day forms of life connect. All the branches below the last fork terminate, representing ancient extinct organisms that have left no living descendants.

One theory suggests that, instead of dying out, some of these earlier forms of life may have joined forces to create new branches on the tree of life. Mitochondria, for example, the organelles that store and release energy in many cells, are believed to be the remains of once free-living organisms that invaded host cells and lived in peaceful coexistence, using their own metabolic and reproductive processes.

Gradually, over many millions of years, a mutually beneficial, or symbiotic, relationship evolved, with both the host cell and the mitochondria losing their ability to function autonomously. The theory is supported by the fact that mitochondria still retain some of their own DNA, the last remaining trace of their past independence. Recent research adds further weight to the argument, suggesting that chloroplasts, thought to have descended from cyanobacteria, may also have started out as bacterial invaders.

The Scientists

The Collectors

'Genius is only a greater aptitude for patience.'

Comte De Buffon (1707–1788)

Only four of these specimens are butterflies, the rest are moths, illustrating the variation within the natural world.

If you ever have the opportunity to take a look behind the scenes at The Natural History Museum, take it. It's an experience you will never forget. For beyond the locked doors, which occasionally interrupt the flow of display cabinets in the public galleries, lies a sprawling labyrinth of corridors and stairs that leads to the heart of one of the most prestigious natural science research institutes in the world. Stepping across the threshold that separates the public face from the less familiar world of museum science, the magnificently decorated galleries disappear, to be replaced by bare, functional passageways, typical of thousands of academic institutions around the world. But first appearances can be deceptive, as a visit to any one of the Museum's extraordinary collections will show.

The Museum's collections are vast. Gathered over 400 years, they include specimens collected by many famous expeditions, from all over the world, and ranging in age from the origins of the Solar System to the present day. At the latest count, they contain 68 million individual items: 5.2 million plants, 55 million animals (including 28 million insects), 9 million fossils, over a quarter of a million minerals and rocks, and more than 2 million manuscripts and pictures. But it isn't so much

Just a fraction of the Museum's 68 million individual specimens.

the size as the scope, the depth of coverage, and the accessibility of the information held within the collections that together make them one of the most important records of natural history available in the world today.

As earlier chapters have shown, the motivations for the continued expansion of the Museum's collections have gradually changed since the days when Sir Hans Sloane accumulated its earliest specimens, books and illustrations. What began as the fashionable preoccupation of an 18th century gentleman has since become a unique scientific record of the Earth's changing biodiversity over the last 400 years. The growth of the collections, triggered by the great voyages of discovery and expeditions of the 18th and 19th centuries, was bound up with the development of Europe and the biological sciences during those times. Studying the natural world assumed an unprecedented urgency, and the desire to profit from commercially viable species and mineral reserves encouraged a flood of specimens from every corner of an ever-expanding empire. The Museum's collections became more 'scientific', evolving from a mishmash of curious objects into an accurately recorded, precisely organized, systematic inventory of specimens.

As the collections have continued to expand throughout the 20th century, the motivation for this growth has changed, as experts have come to appreciate humankind's role in the destruction of the natural world now and in the past. Coal pollution in the Victorian era and the destruction of rainforests in the 20th century are just two examples of the many activities that have damaged the environment and driven thousands of species into extinction.

Nowadays, the focus has shifted towards the protection of these precious biological resources, and a deeper understanding of the rich variety of living organisms that inhabit the Earth. The more experts can learn about the Earth's biological diversity, the better equipped they will be to advise on the management of living resources and the long-term preservation of the environment. This changing perception of nature, together with rapid advances in areas such as molecular biology and information technology, has not only changed what and how contemporary scientists collect, but also has encouraged them to seek new ways of accessing the information stored within the collections and applying what they learn.

The following chapter will look at how the collections are being used as part of a global initiative, to improve understanding of biodiversity, in addition to solving problems in areas like medicine, agriculture and forensic science. But, first, it is important to set the scene, to begin at the very beginning, with the collections, the only permanent record of diversity that we have, and the starting point from which all museum science must begin.

Many of the most famous 18th and 19th century natural historians, including Linnaeus, Darwin, and Wallace, started out as collectors

Richard Dawkins

SIMONYI PROFESSOR OF THE PUBLIC UNDERSTANDING OF SCIENCE AT OXFORD UNIVERSITY

A great museum of natural history puts on display a tiny fraction of its collection, the tip of the iceberg. Like a well-designed shop window, its display cases are a taster for the massed collections that lie behind the scenes in polished wood drawers and murky spirit bottles. I always think this when I walk around the museum. But this leads me to muse on even larger, if more speculative, icebergs. First, the species so far described are a small sample of the species still uncounted in nature. The entire museum collection, basement cabinets and all, is a limited shop window for huge riches still undiscovered around the world.

Then, the animals that exist, or have ever existed on this planet, are themselves a small showcase for the animals that could have existed, or that tried to exist and failed. This is true at the individual level, and at higher levels too. The Darwinian insight is that within each species the individuals that survive are a successful subset of the range of forms that random genetic variation threw up. Living animals are descended from the survivors: they — and, more to the point, their genes — are the tips of ancestral icebergs. When we see a modern animal in its display case we can imagine a vast, shadowy basement full of unsuccessful would-be ancestors standing beneath it. They died and died and died, and the result is what we see on display.

Just a few of the many skeletons hidden behind the scenes in the Museum's osteology storeroom.

That is familiar; it is Darwinism, though it may be a slightly unfamiliar way of putting it. But the train of thought has not yet run out of steam. We can push on to more fanciful regions. The species that exist, the orders and classes and phyla that we see, are a small subset of those that could have existed. To borrow a thought experiment of the evolutionary scientist Stuart Kauffman, if the geological clock could be reset and evolution rerun, not just once but thousands of times independently, how similar, or how different, would be the results? From the natural experiments of Australia, South America and Madagascar, we have some idea of what happens if mammal evolution is rerun. We repeatedly get a range of types, a mammal fauna, which is not identical but in which there is a tantalizing parallelism. Nobody knows what would happen if the clock were reset to the origin of life itself, or to the origin of multicellular life. The question is half-tantamount to one that fascinates us all: is there life on other worlds, and how similar is it to our kind of life?

The museum in which I stand is itself one display case in a gigantic meta-museum, a museum of museums. The fauna that actually exists is the tip of the iceberg in a huge collection of possible faunas, imaginable or perhaps beyond our imagining. It is time to walk out of those huge gothic doors, into the sobering daylight of this planet, today's real world. But I'll be back.

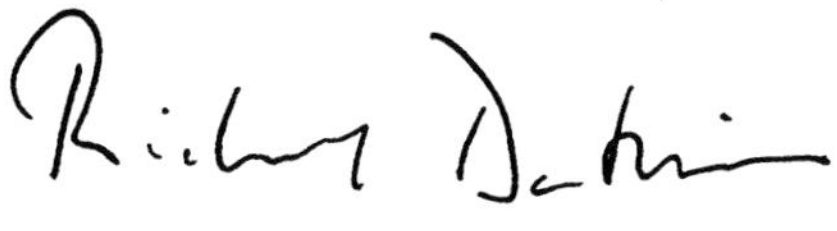

Richard Dawkins

and explorers, tirelessly gathering, preserving and cataloguing at least one of every new species encountered during years of exploration. The majority of these specimens have since become part of the Museum's collections as have countless others, contributed by the many equally dedicated, often courageous, but less well-known men and women who set out to discover 'new' lands for themselves. The stories of a tiny fraction of these collectors appear in the following section. They have been selected more or less at random and have been chosen to illustrate some of the more personal motivations that lie behind the growth of the Museum's collections.

Great Collectors

MARY ANNING (1799–1847)

She sells sea-shells on the sea shore,
The shells she sells are sea-shells, I'm sure
For if she sells sea-shells on the sea-shore,
Then I'm sure she sells sea-shore shells

Terry Sullivan, 1908

The famous American palaeontologist Stephen Jay Gould has described Mary Anning as 'probably the most important unsung (or inadequately sung) collecting force in the history of palaeontology'. Born in Lyme Regis on 21 May 1799, she was the daughter of Richard Anning, a cabinet-maker who supplemented his income selling fossils gathered from the local hills and cliffs. Even as a young girl, Mary accompanied her father and brother Joseph on fossil-collecting expeditions, and together they discovered and assembled the famous 'crocodile', or ichthyosaur, that attracted widespread interest and is now on display in the public galleries of the Museum. Richard Anning died in 1810, when Mary was only 11 years old, leaving his widow and children destitute until the sale of the ichthyosaur in 1815, for the then generous sum of £23, lifted them out of crisis. After her father's death, Mary continued to scour the cliffs and beaches in search of new fossils to sell through the family's tiny shop on Broad Street, Lyme Regis. Over the next 30 years, her uncanny ability to spot 'what others failed to see' earned her the reputation of being one of the most successful, and knowledgeable, fossil collectors in Britain. Her astonishing list of discoveries include several major finds: at least three complete ichthyosaurs, two plesiosaurs, coprolites that she correctly identified as fossil faeces, and at least one pterosaur.

Mary Anning, the pioneer fossil collector from Lyme Regis.

Mary Anning was an exceptional woman. She came from a working-class background, and lacked any formal education or training. Yet, despite these obstacles, her talent for finding specimens and her rare understanding of fossil anatomy earned her the respect of most of the leading palaeontologists of her day. Her long list of influential friends included Louis Agassiz, Sir Henry De la Beche, the Reverend William Buckland and Sir Roderick Murchison. Mary Anning died from breast cancer in Lyme Regis on 9 March 1847, and in 1848 De la Beche paid her a glowing tribute in his presidential address to the Geological Society, describing her 'talents' and 'untiring researches' that contributed 'in no small degree to our knowledge of the great Enalio-Saurians, and other forms of organic life entombed in the vicinity of Lyme Regis'. It was an unprecedented compliment for a woman at that time, but entirely fitting for someone who had already become the stuff of legend, celebrated in poems and cartoons.

An international symposium was held in Lyme Regis in June 1999 to celebrate the bicentenary of Mary Anning's birth, at which Stephen J. Gould, Sir Crispin Tickell (a direct descendent of Mary Anning) and Angela Milner, among others, spoke about her great contribution to science.

MARY HENRIETTA KINGSLEY (1862-1900)

Mary Kingsley was born in London, on 13 November 1862. Her father, George, was both a doctor and a seasoned traveller, who spent long periods away from his family studying anthropology. Like many women of her time, Mary lived a restricted existence. Bound by convention, she devoted most of her first 31 years to nursing her invalid mother and managing the household, leaving her little opportunity to explore the world, except through the books within her father's library. Mary idolized her father, and when he died unexpectedly in his sleep, followed by her mother only six weeks later, her grief encouraged her to pursue an unconventional course of action. Mary decided to travel alone, first to the Canaries, and then, on a more ambitious journey in August 1893 along the coast of West Africa.

Mary's expedition was inspired by a desire to complete the book on primitive religions that her father had left incomplete on his death. But investigating fetishism amongst native tribes was hardly a respectable occupation for a woman travelling unescorted through Africa in the late 19th century. Mary needed another more conventional excuse, and she settled for the study of zoology, seeking the advice of men like Albert Gunther at the British Museum on how best to collect specimens of fish, insects and plants during her stay. Specimen hunting may have been something of a smokescreen to start with, but Mary soon discovered that she had a genuine interest and an aptitude for observing nature.

Mary returned from her first trip to Africa in 1894 and sought the advice of Dr Gunther on the 'miscellaneous specimens' she had gathered during her six months away. He was so impressed by what he saw — particularly the fish collection, which included many new species — that he arranged to supply her with 15 gallons of expensive preserving spirit for her next trip to the little-explored Ogooué River in the Gabon.

Mary Kingsley must have made a memorable impression on the people that she met on her travels. A lone figure, dressed in conventional black mourning, she rejected the notion of wearing the more masculine clothes favoured by most of the other great

Mary Kingsley collected in Africa in the late 19th century.

women travellers of her time. Instead she 'strode, or sloshed through the mangrove swamps in full Victorian skirts, with a hat and an umbrella, as though about to produce a visiting card from her handbag.' (Peter Raby, *Bright Paradise*, 1998).

But if she looked conventional, her behaviour was anything but. She lived with native African tribes, including the Fan, a people notorious for their ferocity and cannibalism. From them she learnt how to paddle and fish from canoes and defend herself from marauding crocodiles with the paddles.

Mary's second journey to Africa lasted nearly two years. During this time, she collected 65 species of fish, including 8 new species, and 18 species of reptiles. One of the lizards that Mary collected in Lambarene was only the second specimen of this species ever to be found. It was particularly prized by Museum staff, who had been hoping to collect one for over ten years.

Unlike most people of her day, Mary tried to understand the native Africans and their culture, rather than judge them according to the customs of white European society. When she returned to England, she campaigned vigorously for improvements in their welfare and treatment by the authorities in the British colonies. By 1900, she was eager to travel again, but this time she chose as her destination South Africa at the time of the Second Boer War. She planned to collect fish for Dr Gunther from the Orange River, but found herself nursing Boer prisoners in a fever camp in Simonstown instead.

Despite drinking wine instead of water and smoking to ward off infection, she died two months later from a heart attack, brought on by a violent bout of fever. She was buried at sea on 3 June 1900 and soon after, the Royal African Society was founded in her honour.

HENRY WALTER BATES (1825–92)

Henry Walter Bates was born in Leicester in 1825. The son of a hosier, he left school at 13, but attended evening classes at the Mechanics Institute and won prizes for Greek and Latin. He also developed a keen interest in insects, which led to the publication of his first scientific paper on beetles, at the age of only 18. Bates served an apprenticeship, first at his father's business and later at a brewery in Burton-on-Trent, but his love of entomology proved too great to allow him to settle down to his chosen career. In 1847, both Bates and Alfred Russel Wallace, who had become regular correspondents through their mutual interest in entomology, read the American W.H. Edward's book *A Voyage up the Amazon*, and were so inspired that they decided to follow in the author's footsteps. A year later, careful planning became firm reality, and they found themselves at the Liverpool docks on board the *Mischief*, a small trading vessel bound for Brazil.

Like Wallace, Bates eked out an existence selling many of the insects and other specimens he collected along the Amazon to dealers in Europe. It was a meagre income, but enough to allow him to

The perils of hunting toucans; this lithograph by Henry Bates is from *The Naturalist on the River Amazons.*

indulge in his passion for natural history. Wallace returned to England after four years, but Bates continued to travel up and down the Amazon for another seven, often sleeping overnight in a cramped canoe, surviving yellow fever, violent storms, and the 'inhospitable wilderness' of the rainforest. He collected hundreds of thousands of specimens, including more than 14,000 insect species, 8000 of which were new to science, and over 7000 species of other animals, including birds, monkeys and anteaters.

Ever conscious of the fragility of his specimens, Bates filled copious notebooks with meticulous observations and detailed pencil and watercolour illustrations of the insects he encountered on his exploration of the Amazon. It was this methodical attention to the smallest detail that led Bates to observe the type of mimicry that came to bear the name 'Batesian mimicry' in his honour. He noticed that certain species of harmless Brazilian butterflies avoided falling prey to birds by appearing and behaving very like species that tasted unpleasant. In 1862, Bates presented his observations to the Linnean Society in London, and provided a delighted Darwin with an important piece of evidence to support his theory of natural selection.

In 1863, with Darwin's support and encouragement, Bates published an account of his travels, called *The Naturalist on the River Amazons*. The book was a great scientific and popular success, and Bates went on to be elected to a number of prestigious posts, including president of the Entomological Society in 1869 and 1878. He was made a fellow of the Royal Society in 1881, and died eleven years later, aged 67, from bronchitis and influenza.

LIONEL WALTER, 2ND BARON ROTHSCHILD (1868–1937)

Lionel Walter Rothschild, the eldest son of wealthy financier Nathaniel Mayer Rothschild (1840–1915), became interested in zoology as a young boy. His enthusiasm was fuelled by a meeting with Dr Albert Gunther, the British Museum's Keeper of Zoology, who happened to chance upon the 13-year-old staring intently at the zoological exhibits soon to be moved from Bloomsbury to South Kensington. Gunther became an important influence in Rothschild's early years, and encouraged him to spend a short time studying in Germany before entering Cambridge as an undergraduate to study natural science.

The giant moa, *Dinornis novaezealandiae* (originally *Dinornis ingens*), from Lionel Rothschild's book *Extinct birds*, 1907.

Lord Rothschild, in his buggy pulled by a zebra, must have made an impressive addition to the traffic near his private museum at Tring.

Rothschild's privileged background allowed him to collect and buy many zoological specimens from an early age, rapidly accumulating butterflies, moths, birds, and an impressive library that soon outgrew their allotted sheds and hired rooms in Tring, Hertfordshire, where the family had its private estate. His father provided a generous solution to the problem. For his son's coming of age, Lord Rothschild granted him a piece of land on the outskirts of the family home to build a private museum to house his expanding collection.

Walter Rothschild was well known for his eccentricities. He often drove himself about in a buggy pulled by a team that included three zebras, and in May 1920 caused a sensation when he rode through London, in an open-topped taxi, accompanied by a 2.85 m high model reconstruction of a giant moa. Rothschild commissioned the model of the extinct New Zealand bird to exhibit at a meeting of the British Ornithological Club held at Pagani's restaurant in Great Portland Street.

Rothschild lacked the business acumen that had made his family so wealthy, and after his father's death, his passion for collecting, researching and displaying natural history specimens in his private museum brought him close to financial ruin on several occasions. Eventually, mounting debts forced Lord Rothschild to sell parts of his collection and in 1931 he sold his entire bird collection to the Museum of Natural History in New York.

The loss of Rothschild's treasured birds played heavily on his mind, and he became an invalid in the final years of his life, after breaking his leg in an accident. He died of cancer on 27 August 1937, having bequeathed his museum at Tring, its buildings and all the remaining collections to the Trustees of the British Museum (Natural History). Walter Rothschild collected over two million mammals, birds, reptiles, amphibians and invertebrates during his lifetime. The bulk of these specimens remain with The Natural History Museum, a lasting reminder of the generosity of a Victorian collector who dedicated his life and income to improve our understanding of the natural world.

What do the Collections contain?

Traditional collections were based on whatever was most easily preserved — wood, fruit, seeds, bones, whole organisms preserved in spirit, dried plants in herbaria and animal skins — together with a library of information, drawings, detailed descriptions and photographs. But these requirements have changed, as scientific and technological advances have presented taxonomists with a growing number of new techniques, not only for preserving specimens but also for differentiating between species, and understanding their evolutionary relationships.

The invention of the light microscope, for example, allowed taxonomists to study

The chiffchaff, *Phylloscopus collybita*.

single-celled protists in detail, and other scientists to look at the differences in the microstructure of crystals and fossils for the first time. The transmission electron microscope (TEM) revealed greater detail, enabling researchers to distinguish subcellular characters that led to further advances in the classification of the protists. The scanning electron microscope (SEM) yielded more surface detail, revealing minute differences in the structure of minerals, and of individual diatoms, foraminiferans and pollen grains.

Storing microscopic specimens on slides is only one of the ways in which the collections have expanded to accommodate new ways of analysing the natural world. Over 200 years ago, Gilbert White (1720–1793), Vicar of Selborne, Hampshire, and one of the greatest of all British naturalists, helped to popularize the study of nature when he described his observations of local wildlife in his classic book *The Natural History of Selborne*. White was the first naturalist to distinguish three species of leaf warbler in Britain, two of which appeared virtually identical: he discovered that *Phylloscopus trochilus*, the willow warbler, and *P. collybita*, the chiffchaff, could be distinguished by their song. Nowadays, songs can be recorded, digitized and analysed by computer to show the degree of similarity between different species, while other important behavioural differences, such as mating displays, can be captured on video film.

RECENT ADVANCES

Over the past twenty years, advances in molecular genetics have added yet another new dimension to contemporary collections. Nowadays, there are walk-in freezers that store cell and tissue cultures, germ plasm, DNA banks and other molecular records, all of which

LEFT: **Chrysotile asbestos fibres seen under the optical microscope. Each 'fibre' is actually a bundle of thinner fibres.**

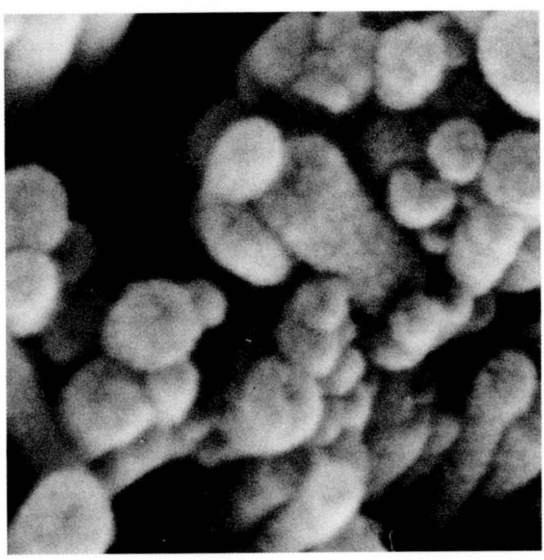

LEFT: **High resolution SEM image of chrysotile asbestos fibres. Individual fibres can just be seen to be tubular in shape.**

LEFT: **This TEM image shows chrysotile asbestos fibres are made up of curved silicate sheets. The sheets, which are 0.73 nm apart, form tubes. Under certain conditions, 15 dark radial spokes appear in the image.**

LEFT: **Collectors in the Amazon, 1873.**

RIGHT: **Disturbance to habitats is minimised when collecting plants. Botanists are studying *Asplenium csikii*, an extremely rare rock fern found in the British Isles and other parts of Europe. All of the information needed to assess the variation and distribution within and between 23 different populations has been gathered by removing only a single frond from *c.* 500 plants.**

are providing taxonomists with new ways of investigating the evolutionary relationships between different organisms. Experts hope that cryopreservation (the preservation of living cells by freezing) may also offer a means of conserving the genetic potential of so far unidentified microorganisms in samples taken from disappearing habitats.

Collecting and Sampling

Although the Museum continues to acquire most of its specimens through collecting and sampling, methods and practises have changed considerably since the days of the great Victorian collectors. The 19th century collectors often travelled alone, for years at a time.

Nowadays, most collecting is associated with research projects and relies on the collaboration of teams of scientists working towards a common goal. Every year, for example, botanists from The Natural History Museum team up with local scientists from all over the world to collect plants for a specific purpose. They might be needed as tools to identify and monitor biodiversity, or to understand and map the evolutionary relationships among certain species.

Contemporary research is more focused on single species and, as the following chapter will show, experts have largely abandoned the Victorian method of collecting one of each species in favour of gathering much more detailed information about species across their range.

CONVENTION ON INTERNATIONAL TRADE IN ENDANGERED SPECIES

If you live in the UK, before you go on your next holiday the Department of the Environment recommends you read their guide on CITES, the international treaty that covers more than 21,000 species of endangered plants and 3000 species of endangered animals. One of the most successful treaties in the conservation of global wildlife, CITES makes the trade of these plants and animals, or their body parts, illegal.

The endangered cycad tree (male), *Cycas media*, with its cone.

INTERNATIONAL LEGISLATION

Recognition of international laws, set up by the various professional bodies that govern which biological and geological specimens may and may not be collected, is another aspect of collecting that has changed substantially over recent years. Whereas naturalists were once free to roam abroad, gathering specimens without recourse to authority, now they must abide by the laws and regulations of the countries in which they collect.

Museum scientists are also governed by their own code of practise, designed to minimize damage to the environment and to encourage other countries to develop their own national inventory of biological and geological specimens. Where possible, Museum experts prefer to collaborate with a local institute or recognized organization on fieldwork conducted outside the UK. Specimens are collected only as part of an agreed research project, or an environmental impact assessment, and no more plants and animals must be collected from their habitat than is strictly necessary for essential research purposes. Protected species must not be collected without prior authorization of the country concerned, and then only if certain conditions apply.

Once the research project is complete, the first set of specimens is returned to their country of origin, so that local scientists and conservation initiatives have first-hand access to important information that will help them manage and preserve their biological resources. The Natural History Museum keeps any duplicate specimens, adding valuable new information to its collections, and providing an important safeguard against the loss of important data should the first set of specimens deteriorate. This may not be an important issue for countries that have ample funding and that are able to afford suitable storage facilities, but for some of the Third World nations with limited resources, and often hot, humid climates, the difficulties of preserving specimens can be a matter of key concern.

LIFE IN THE FIELD

The motivations for collecting and the code of practise might have evolved over the last four centuries, but in some respects the experience of being out in the field hasn't changed a great deal from the days of 18th century collectors, such as Linnaeus. Learning more about parts of the world about which we know very little inevitably points to remote locations far from the beaten track, and the comfort of even the most modest accommodation. In the words of palaeontologist Richard Fortey: 'There are usually no hotels where the best fossils are found. Instead there are tents and shacks or sleeping out under the stars'.

TO CATCH A BEETLE

Collecting plants and animals that live in inaccessible places requires a special blend of ingenuity, patience and courage. Martin Brendell, beetle curator, describes the art of collecting insects that live more than 30 m above the ground in the tree canopy of a tropical rain forest:

'When trying to discover the diversity of insect life living in a single giant forest tree, the aim is to collect a set of samples, over the shortest period of time, that are as complete a representation of the insect fauna of the tree at that 'moment'. At the same time it is vital to minimize the impact of our investigation on the tree itself and on the surrounding area. Climbing is one of the ways of finding out about the insects that live within the foliage and along the branches, but it can be extremely dangerous, and in the rainforest, with its open canopy, not the most efficient way to sample an entire tree. In the old days, entomologists paid to have trees chopped down so that they could race forwards the moment the branches hit the ground, and grab as many insects as possible. However, in more recent times, we have developed a variety of less destructive and more effective techniques.

Today, in order to get accurate information about the diversity of insects that live in the forests, we need to sample whole insect communities of individual tree canopies, but this doesn't mean that we need to sample every tree. We may decide to target five separate species of tree in a particular forest. Far from tackling them at random, we select one individual from each species and extrapolate the information gathered from each tree to build up a picture of the entire forest. By hauling a mechanical fogger, loaded with knock-down insecticide, high up into the tree canopy, we are able to target, quite precisely, the whole of the tree we are interested in, and collect nearly all of the insects in one hit. The fogger emits an insecticidal fog, activated by remote control, that drifts throughout the canopy of the chosen tree. We control the environmental impact of the insecticide by using a biodegradable formulation that is destroyed by sunlight. The dazed insects fall from the tree into catchment trays, hung on ropes at ground level, far below, and are gathered into bottles of preserving fluid that usually consist of ethyl alcohol.

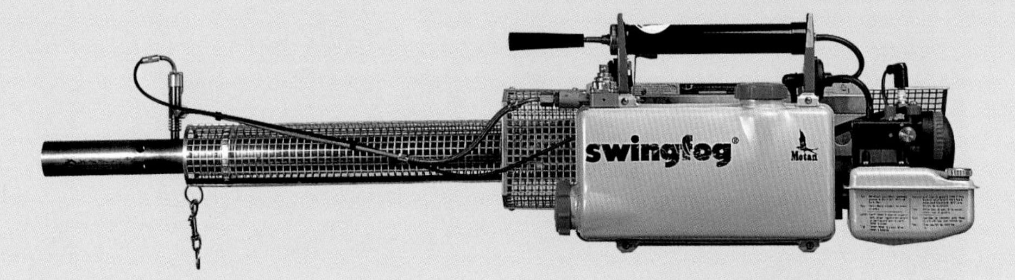

Mechanical fogging device.

Although we try to reach every insect, inevitably some will escape to recover, unharmed, from the effects of the insecticide. When we come to sort the trays of insects we sort the individuals into 'morphospecies' according to their bodyshape, and each of these is given a code number. Inevitably it isn't possible to name every morphospecies in the

Catchment trays for collecting insects from fogging.

The Peter Hammond flight intercept trap.

field, because some will be new to science and will have to be described and given scientific names later. On average, we will sort up to 10,000 individuals from each tree, which will include several hundred different species.

Passive methods of sampling include traps and lures. The Peter Hammond flight intercept trap consists of a 2.4 m by 0.9 m black nylon mesh that is set up in the flight path of passing forest insects. The insects literally crash into the 'invisible' mesh, and fall into trays of liquid that contain a mixture of water and chloral hydrate crystals, which kills and preserves them. Using this trap, we can collect about 1000 beetles per night, which may include 100 different species. There are also smaller, more portable traps like the Malaise trap, a tent-like funnel that lures insects towards the sunlight and into a bottle of preserving liquid fixed at the top.

These traps can be mounted on frames and hoisted up into the trees, which allows us to map each species in relation to its precise location within the tree canopy. We can enhance the effectiveness of other traps by using brightly coloured yellow trays that mimic flowers or, in the case of the Tullgren funnel, by warming the funnel with a light to encourage insects in leaf litter samples to move downwards into a liquid trap that awaits them below.

Understanding the behaviour and life cycle of certain species also helps us to outwit our quarry. Beetles have colonized every conceivable terrestrial environment on earth and,in the case of certain specialist species, we have to employ other methods, besides traps, to catch them. The site of a road-kill may not inspire many people, but to me it means the possibility of discovering new species of beetles that inhabit rotting carcasses! Entomologists are always searching for new ways of accessing insects that dwell in unusual places.

I remember once plotting to capture specimens of the specialist beetles that live deep within subterranean ant nests in Central America. I settled on gently inserting some dried grass attached to a length of string, and left it pushed deep down into the nest for a day or two, in the hope that the beetles would crawl into what might appear to be a comfortable new bed. When I returned to remove the grass, I was thrilled to discover several different species of beetle nestled inside my grass ball.'

Tullgren funnel.

Palaeontologists sacrificing their home comforts to work in the field.

Getting lost, lack of sleep, adjusting to drastic changes of diet and temperature, and coping with the dangers posed by the local wildlife are only a few of the challenges that modern explorers continue to share with their 17th, 18th and 19th century predecessors. Scientists delight in sharing their favourite expedition stories, from eating sea slugs and goldfish in China to walking through the wilderness to an isolated lake, only to discover that they have left their two weeks supply of teabags behind in the jeep they abandoned twelve hours earlier.

PRIVATE COLLECTORS

While the Museum gathers most of its new specimens through collecting by its own staff, it also receives them from private collectors and dealers, as well as operating a system of exchanges with other museums and institutions. Oliver Crimmen, curator of the fish collection, remembers one donation in particular:

'We had a phone call from a representative of the food hall at Harrods to say they were laying out a whole basking shark to be sold in slices and would The Natural History Museum be interested in keeping the head? Transporting a large specimen like a shark over any distance tends to be an expensive operation and the close proximity of the Harrods store meant a rare opportunity to get hold of a large head. There is currently only one known species of basking shark in the world, the Museum didn't have a specimen, and because the head contains vital anatomical information we were delighted to accept the offer. But how could we transport our prize before Harrods was obliged to throw it away? The solution couldn't have been simpler. Grabbing a specimen barrow, I walked a brisk half-mile to the store before the food hall could change its mind. If pushing an empty wheelbarrow through Knightsbridge raised a few eyebrows during my trip to Harrods, nothing could prepare Londoners for the sight of my return to the Museum. A triumphant curator wheeling the head of a basking shark through the lunchtime traffic? Even the taxi drivers looked surprised!'

Four and a half miles of shelving house specimens preserved in jars of ethyl alcohol.

CONSERVATION

The conservation of millions of specimens housed within the Museum's collections is a huge but essential task in the quest to preserve this unique record of the earth's changing history. Many of us might assume that the staff who work on the collections spend most of their time working with new specimens, but in reality much of their time is devoted to the maintenance of specimens that may have been a part of the collections for hundreds of years.

In the Botany Department, for example, four people work full-time, not only mounting new plants for the Herbarium (the Museum's collection of plants) but also repairing and remounting old specimens that are in danger of disintegration or damage.

Shirley Brennan, who has been mounting plants for over ten years, recalls: 'I spent my first two years learning how to mount new specimens. There are many different types of flowering plants and you have to learn how best to display them so that the scientists can access the areas that are important to them for their research.

After this initial training, I moved on to the more delicate work of repairing and remounting fragile specimens from the Herbarium. Once again, it took considerable practice before I mastered these skills and was deemed sufficiently expert to work on the renovation of the department's critically important type specimens. It's an unusual job, but I've thoroughly enjoyed my years at the Museum, especially the time that I was able to go home and tell my son-in-law, a policeman, that I had just remounted its collection of cannabis plants!'

A bromeliad, *Guzmania lingulata*, collected by Sir Hans Sloane from the voyage to Jamaica, 1687–1689.

What Happens to the Specimens After They are Collected?

PREPARATION

Back in the Museum, the specimens must be prepared and labelled ready for identification, and protected to prevent disintegration or attack from pests. The staff employ a huge range of preparation techniques that vary from collection to collection: fossils are extracted from surrounding layers of rock matrix, for instance, plants are pressed and mounted, insects pinned, fish dissected, and slides made of microscopic organisms.

All the work on the collections, and the collections themselves, are housed in rooms where the temperature and humidity are carefully monitored to guard against deterioration. Freezers kill pests on incoming specimens, a far less hazardous method than the noxious poisons that were used to preserve them in the past.

Some of the collections are still housed in the original wooden cabinets and drawers designed for the Museum that date from around the 1770s, but they are gradually being replaced by modern compactor shelving, which saves on valuable storage space and offers protection against fire damage.

IDENTIFICATION

Once the specimens have been prepared, the next task is identification. Making a correct identification isn't necessarily as straightforward as it sounds, and errors can have far-reaching implications. In the 1970s, for example, a mealybug infestation spread through tropical Africa from the Congo, threatening to destroy cassava, the primary source of food for more than 200 million people. Controlling the outbreak using conventional pesticides was too expensive, and a biological control programme was set up instead. The mealybug responsible for the crop damage was thought to have originated in either Central America or northern South America, and parasitic wasps from this region were tested against bugs from both the Congo and Zaire.

The tests were unsuccessful, and it was only when an entomologist visiting Paraguay sent a sample mealybug back to the UK for identification that taxonomists from the Centre of Agriculture and Biosciences International (CAB International), working with colleagues at the Museum, realized that there were, in fact, two closely related mealybugs damaging cassava in Africa. They had correctly identified the region from which one of the species had originated, but realized that the other had arrived from central South America. The search for an effective parasite switched to Paraguay, Bolivia and southern Brazil.

Out of 30 natural enemies of the South American mealybug, one species, a minute parasitic wasp, *Apoanagyrus lopezi,* proved spectacularly successful at killing it. Introduced in 27 African countries, the wasp spread rapidly, populating an area of approximately 3 million square kilometres. The mealybug is now successfully under control and, although the research programme cost about 10 million $US, experts believe that as well as the prevention of further human suffering, the introduction of *Apoanagyrus lopezi* has meant a huge financial saving for Africa of more than 160 million $US per year.

This flightless species of darkling beetle, *Amianthus bufo,* was one of the zoological specimens collected by David Livingstone during his travels along the Zambezi.

Identification can take a long time. If you visit any museum collection there is always a backlog of unsorted specimens, pinned in trays and filed away into drawers, waiting to be identified and described. In some cases, they have been waiting for years. Some of the Museum's unidentified beetles date back to the late 18th century. There are specimens from Captain Cook's voyages and Darwin's travels on the *Beagle*, and entomologist Martin Brendell recently identified a beetle collected by the Scottish explorer David Livingstone (1813–73) on his travels along the Zambezi river in Central Africa. Martin has identified it as belonging to the darkling beetle family, Tenebrionidae, appropriately named from the Greek meaning dark and gloomy. The much more familiar cellar beetle ('black beetle'), flour beetles, mealworm beetle and relatives also belong to this family.

Why does this backlog occur? In part, it is a question of time: identification is only one of the many tasks that Museum curators, the scientists who manage the collections, have to juggle on a daily basis, but it is also a reflection of the sheer number and diversity of specimens concerned. Often, the curators lack the expertise or the literature that they need to make a successful identification. This is why the system of making specimens freely available on

TICODENDRON

In 1989, botanists working in Costa Rica were reminded that plants aren't always what they appear to be. They had been collaborating with international scientists for eight years to identify and database the many thousands of species of plants that grow in Costa Rica for *Flora Mesoamericana*, a guide to the identity and distribution of flowering plants and ferns in central South America, but nobody paid much attention to the common alderlike tree that grew everywhere, offering much-needed shade from the sun. Nobody, that was, until Jorge Gomez-Laurito and Luis D. Gomez decided to take a closer look, and realized that it wasn't an alder at all. Some of the leading systematic botanists in the world had been blissfully unaware that they were strolling past, leaning against and sitting under a new species of tree.

Ticodendron.

loan to visiting experts, or sending them out to other academic institutes, is such an important service. External experts request these specimens to help them with their research, and receive them on the understanding that they will identify, label and describe them before returning them to the Museum.

About 64,000 specimens are sent out on loan by the Museum every year. In the Entomology collection this includes 600–700 new loans of 24,000–38,000 specimens, made to 300 borrowers in more than 40 countries. This makes the management of this system in itself a time-consuming task.

THE IMPORTANCE OF TYPE SPECIMENS

If you were to look through the thousands of drawers and containers that house the collections of the Museum, among the first things you would notice are the markers on the labels that indicate 'type specimens'. When a detailed examination of a specimen reveals that it belongs to a new species, it is designated a type specimen. This is extremely important, because it is the physical record upon which the published name and description for that species is based — in effect, a blueprint for identifying other members of the same species.

Thousands of new species were discovered as a result of the great voyages of discovery in the 18th and 19th centuries, which is why the Museum has such a rich source of type specimens gathered from every corner of the globe. But it is still remarkable to think that experts seeking to determine the identity of a particular plant, animal, fossil or mineral often refer to type specimens collected and neatly labelled by some of history's most famous naturalists, including Darwin, Linnaeus, Hooker and Huxley.

Sadly, many type specimens, such as the majority of those in Alfred Wallace's Brazilian collection destroyed in a ship fire in 1852, were either lost or perished during the long voyages at sea in the 18th and 19th centuries, but often drawings or paintings of the original plant or animal still survive. In these circumstances an illustration, the closest approximation to the real thing, replaces the lost type specimen and becomes the 'iconotype' for that species.

Artists and illustrators

The reading room that houses the Museum's collection of rare books and prints is a treasure trove of original sketches, watercolour drawings, oil paintings and prints accumulated from over four centuries of exploration and discovery. Containing over half a million works of art, it is the largest collection of natural history illustrations in the world, made all the more remarkable because Museum scientists consider most of them a vital tool in their taxonomic work. Hundreds of men and women have contributed to this collection, many without any formal training, but every one of them has produced a lasting record of plant and animal species that continues to help experts identify new species and provide valuable information about the changing diversity of life on earth.

The following selection represents only a tiny fraction of the artists who have contributed to the Museum's collection. They have been chosen to illustrate the most common motives that encouraged these remarkable men and women to observe, paint and draw.

THE PROFESSIONALS

A growing fascination with the natural world, fuelled by tantalizing reports of exotic plants and animals from faraway continents and the few living and preserved specimens that survived the long voyages at sea, created many new opportunities for professional artists in the 18th and 19th centuries. There was a huge demand for information and before the invention of photography, high-speed travel and television, paintings and drawings were one of the most effective ways of conveying it.

The English botanist Sir Joseph Banks (1743–1820) pioneered the tradition of including natural history artists on the great voyages of discovery. These travelling artists occupied their time sketching fresh specimens in the field and making a detailed record of their colours so that they, or other professional artists, could work the initial sketches into finished paintings back home. Having already established his reputation as an exemplary botanical artist in 1801, the Austrian Ferdinand Lucas Bauer (1760–1826), regarded by many as the greatest natural history artist ever, was selected by Joseph Banks to sail on HMS *Investigator* on an expedition to chart the coasts of Australia. Experienced and extremely hard-working, Bauer made 1770 pencil drawings of plants and 303 drawings of animals, many of which he later turned into stunningly realistic watercolours that reflect the rich diversity of Australian flora and fauna.

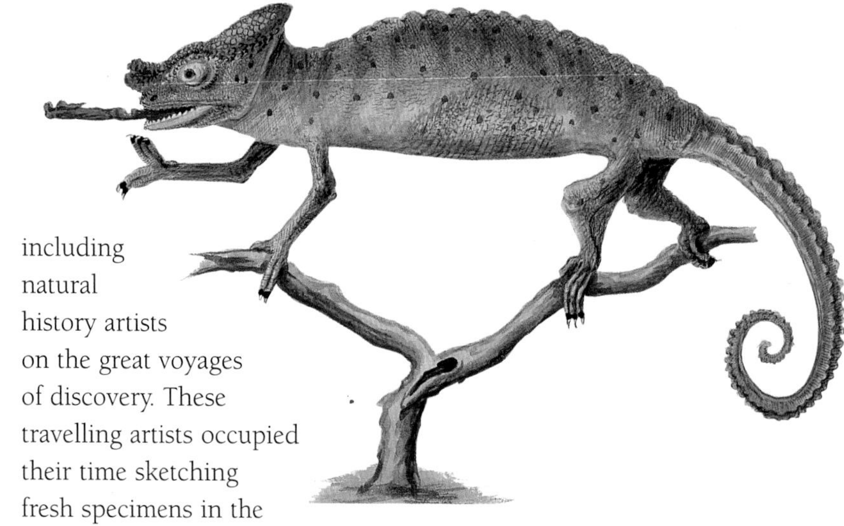

Chameleon by Sarah Stone. Sarah Stone's skills in recording the minutest details of the specimens that she painted are evident but some of her animals appear less than lifelike. This was more a reflection of the skills of the taxidermists than lack of artistic talent. The specimens had often been stuffed for display with very little understanding of the underlying musculature.

Wombat, *Vombatus ursinus*, by Ferdinand Bauer.

Steve Jones

PROFESSOR OF GENETICS, UNIVERSITY COLLEGE, LONDON

I have spent many summers studying the biology of snails and slugs. The animals are remarkable enough, and their genes more so. Some snails carry inherited cues of identity on their shells in the form of distinct patterns of colour and banding; this is useful for trying to understand why genetic variation is there, and why different populations differ so greatly in the genes they carry. Many snails and slugs are hermaphrodites — with each individual both male and female — and some can choose to fertilize themselves or not, more or less at will; and this, too, is a useful tool for those interested in the evolution of sex (which is, after all, what genetics is all about).

Tens of thousands of specimens of 'my' snails are hidden in the depths of the Museum, away from the public gaze. I am slightly ashamed to say that the first time I visited, in the mid-1960s, was to look at the snail collection and make sure that I could identify the various kinds on which I hoped to work (although, I must say, I had no thought that I would still be studying them thirty-five years on). Like most of those who venture behind the scenes, I was more impressed by what was hidden away than what was on view (although the Museum's public face has become more friendly than it was in those days).

Over the next three decades there was plenty of science, but in many ways I learned more about people. In Bosnia in the sixties I was baffled by the warnings from one set of villagers that we would be mad to go on to the next village (and, a few years ago, I saw the village in which we had our base in flames on the TV news). In Aberdeenshire, in the early 1980s, an old man who asked me what I was doing as I was picking snails up on the dunes replied 'O me, me; these are tairrible times!', as he thought I was unemployed and gathering them for food. In Reagan's America, I got a ticket (and a minor police record) for collecting molluscs without a permit.

What excited me most on my first visit to the Museum was the continuity with the past; some of the labels identifying the specimens were (or so the curator claimed) in Darwin's handwriting and, of course, there were links with the far more ancient past in the millions of fossil snails in the collection. What excites me now is the thought that my own snails, in their own very modest way, are hidden away among the collections and that — some day, say twenty years from now — a young biologist will come to study them and come up with a new idea: and that, perhaps, that young biologist is already among the hundreds of excited children who fill the Museum's halls every day.

Banding patterns on *Cepaea nemoralis* snails vary between different populations.

J. S. Jones

Crowned pigeon, *Goura cristata*, by Sarah Stone.

Sarah Stone (1760–1844) provides a fascinating example of one of the many natural history artists who made their living not by travelling, but by drawing and painting the specimens and artefacts brought back to Britain by survey ships towards the end of the 18th century. She was discovered as a teenager by the entrepreneur Sir Ashton Lever, painting zoological specimens displayed in his privately owned museum, in Leicester Square, London. Lever was so impressed by Sarah's work that he commissioned her to record the entire contents of the Leverian Museum, and the result was over a thousand highly detailed water-colours of birds, mammals, fishes, insects, shells, minerals and ethno-graphical artefacts.

Sarah Stone was the first artist to paint many of the specimens that arrived from Australasia and the South Seas in the 1770s, 1780s and 1790s. By 1781, her illustrations had become so popular, and were held in such high regard, that she was invited to exhibit at the Royal Academy. At the age of only 21, an invitation to exhibit at the Royal Academy is one of the highest accolades that could be awarded to any artist, and Sarah's youth, and the fact that she was a woman, only serve to make her achievement all the more remarkable.

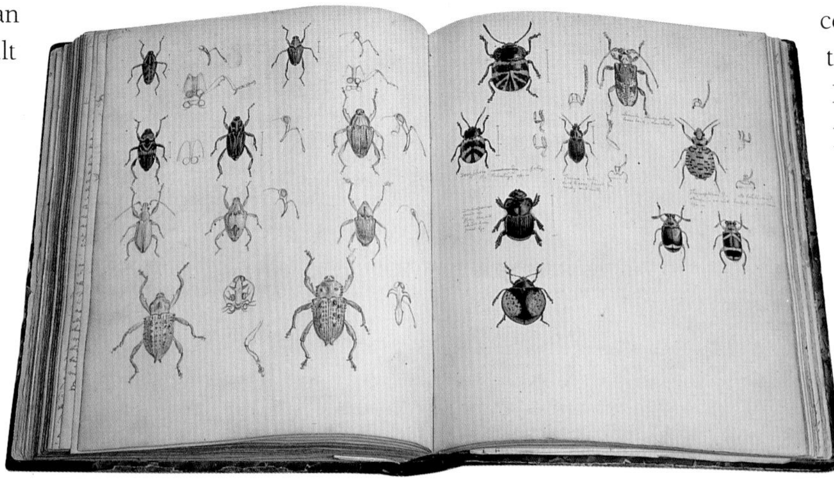

A double-page spread showing coloured sketches of beetles from Henry Bates's journal *The Amazon Expedition*.

SCIENTIFIC NECESSITY

Collecting abroad in hostile environments made the preservation of specimens extremely difficult, and, often, self-taught naturalists such as Henry Bates (1825–92) were encouraged to draw and paint their subjects as a safeguard against the loss of precious scientific information. Many of the stunningly detailed pencil drawings Bates made in his Amazon expedition journal were reproduced in watercolour, to record the precise colours of the specimens under observation.

A FORM OF ADVERTISEMENT

Supplying specimens to fill the curiosity cabinets of the wealthy was a competitive business towards the end of the 18th century, and professional collector John Abbot (1751–1840) painted his specimens to generate orders. Now considered of great scientific importance, Abbot's drawings fill 18 volumes, and are essentially a catalogue of North American insects and birds that he supplied to customers through his agent, John Francillon, in London.

PAINTING FOR PLEASURE

'There is no better way of loving nature than through art.'

Oscar Wilde

One of the best-kept secrets of the Museum's rare books collection are the many exquisite illustrations hidden away in the sketchbooks, notebooks, sheets and letters of unknown amateur artists. They may have lacked the formal training of many of the leading natural history artists of their day, but their enthusiasm and sometimes novel, highly personal interpretation of nature can't help but captivate and intrigue. Often donated by surviving members of the artist's family, many provide a fascinating insight into the lives of men and women who travelled for reasons unconnected with science, and yet whose love of sketching and painting has provided an important contribution to our understanding of the natural world.

Olivia Fanny Tonge's father was a naval office, keen naturalist and a painter who failed to recognize that his daughter's inability to paint landscapes was a symptom of myopia rather than an indication of any lack of talent! Born in Llandilotalybont, Glamorgan, Olivia married in 1878 and had two daughters. She began to pursue her interest in art and design only after the early death of her husband several years later. Between 1908 and 1910, and from 1912 to 1913, she lived in India, where she completed the series of sixteen sketchbooks that were donated to the Museum in 1952, after her death.

Rattan fruit by Margaret **Bushby Lascelles** Cockburn.

The sphinx moth, *Lapara coniferarum*, with a long-leaf pine, *Pinus palustris* by John Abbot.

More than a Collection of Scientific Data

The Natural History Museum is committed to sharing its collections and expertise with as broad an audience as possible. Recently, this led to an exciting project with leading Belgian artist Jan Fabre, 'The Cultural Ambassador of Flanders', organized in close collaboration with the science-art agency, The Arts Catalyst.

Jan is the great-grandson of the famous 19th century French entomologist Jean Henri Fabre (1823–1915), author of best-selling titles like *The Life of the Fly*, and many of his drawings, sculptures and installations are inspired by his great-grandfather's passion for insects. In 1997, for example, *The Lime-Twig Man* at the Arnolfini

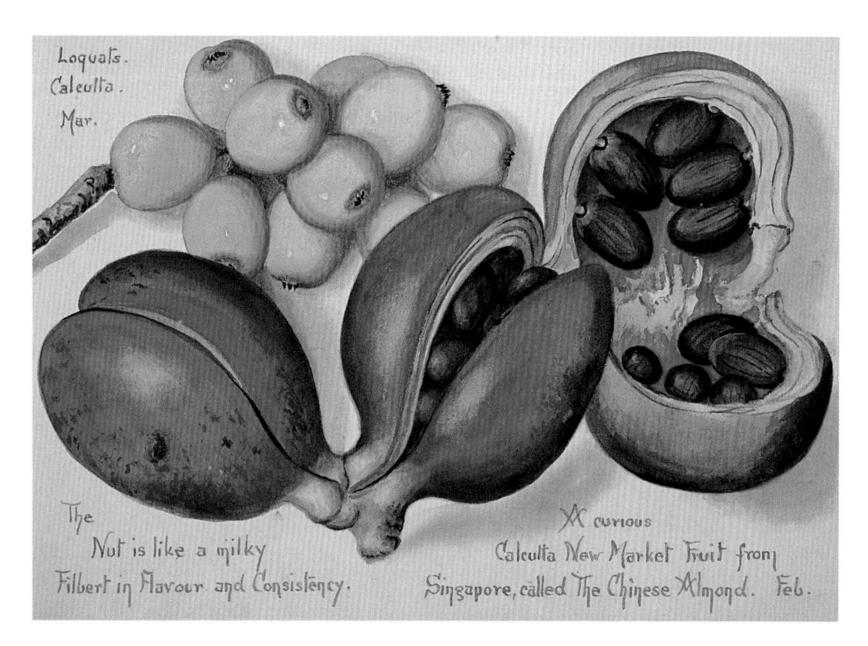

Chinese almond and loquat watercolour by Olivia Fanny Tonge, taken from one of sixteen sketch books presented to the Museum in 1952.

The 'Beekeeper', made from beetle carapaces by Jan Fabre, 1998.

in Bristol included an installation of shrouded figures made from thousands of jewel-like beetle carapaces. And another work, *The Grave of the Unknown Computer*, consisted of multiple crosses, each bearing the name of a different species of beetle written in biro ink.

The Meeting is a filmed conversation between Fabre and five respected entomologists from the Museum, including keeper Dick Vane Wright, associate keeper Rory Post, Ian Gauld, Martin Brendell and Martin Hall, dressed in stunning insect costumes designed by the artist. The film is an exploration of the scientists' and the artist's perception of insect behaviour and society, and the connections that can be made with human civilization. Both the film and the costumes formed part of an installation in the Museum's North Hall in November 1999.

Collections and Controversy: Hoaxes, Frauds and Contra-frauds

People love a good story, and what better than a controversy among learned academics about the interpretation of mysterious remains that hint of mischief or foul play? It's an irresistible combination and, although only a tiny number of hoaxes and frauds have been uncovered throughout the long history of the Museum's collections, it's a topic that will continue to loom large in the public imagination.

THE STORY OF PILTDOWN MAN

On 5 December 1912, the respected scientific journal *Nature* published a report announcing the discovery of fragments of a human skull and a slender ape–like mandible in a gravel pit near Piltdown Common in East Sussex, England. Found by solicitor and amateur geologist Charles Dawson (1864–1916), the remains were thought to date from the early Pleistocene, and the English geologist Arthur Smith Woodward (1864–1944), the then Keeper of Geology, was convinced that they represented the 'missing link' that proved that man had evolved from the apes.

In 1907, the unusually thick, fossilized jaw of an ancient man was found in Maur, near Heidelberg in Germany, sparking a vigorous debate about the evolution of the first humans in Europe. The massive ape like jaw, with its humanlike teeth, dated from the interglacial period of comparatively warm climate that

FASHION AND THE NATURAL WORLD

This unattributed 19th century engraving of Zarafa celebrates her arrival in Paris.

When Zarafa, the first giraffe ever seen in France, arrived in Paris in 1827, she caused a sensation. A gift from Muhammad Ali, Ottoman Viceroy of Egypt, to King Charles X of France, she was shipped to Marseilles and walked to Paris, to the amazement and delight of the crowds that gathered at every resting stage of her 550-mile journey. When Zarafa finally reached Paris, she was installed in the Jardin du Roi, the king's zoological gardens, where tens of thousands of visitors rushed to see her.

Paris went giraffe-mad. Women took to wearing their hair à la girafe, piled so high on their heads they had to abandon the seats of their carriages in favour of the floor. From fabric to crockery, and furniture to food, Paris was awash with products that were either decorated with, or influenced by, the shape of this long-necked, high-legged beauty, and the remarkable cream-and-amber patterns of her coat.

Fashion hasn't always treated the animals it celebrates as kindly as during the Zarafa craze that swept through Paris. In the 1890s, for example, Sir William Henry Flower, the first director of natural history at the British Museum, mounted a campaign to explain how the vogue for feathered hats was spearheading the relentless slaughter of thousands of birds in Britain and many other parts of the world. He was particularly upset by the plight of the snowy egret, the white heron common in Florida, hunted without mercy during its breeding season. The adult egrets made easy targets when they returned to their nests to feed their young, and the orphaned hatchlings were left to starve or perish from hypothermia.

In a letter to the *Times* on 25 June 1896, Sir William wrote 'I have recently noticed many of the gentlest and most kind-hearted among my lady friends, including some who are members of the Society for the Protection of Birds, and who I am sure would never knowingly do any injury to living creatures, adorned with these very plumes the purveyors of female raiment, to salve the consciences of their customers, have invented and widely propagated a monstrous fiction, and are everywhere selling the real feathers warranted as artificial!'

In 1908, under the guidance of various members of staff, Lord Avebury, a trustee of the Museum, introduced a bill to the House of Lords to prevent the import of feathers and bird skins from abroad. It was passed by the Lords, but rejected by the House of Commons. The feather industry was extremely lucrative and, despite another attempt to introduce the Importation of Plumage (Prohibition) Bill in 1914, it wasn't until 1921 that the ban to halt the cruelty that was threatening to drive many species of birds into extinction was finally approved.

***Megazopherus*, flightless beetles from Mexico, are captured, encrusted with precious gems, and worn as living brooches at the end of a chain. Considered by some an attractive item of jewellery, *Megazopherus* only require small amounts of water and food to survive.**

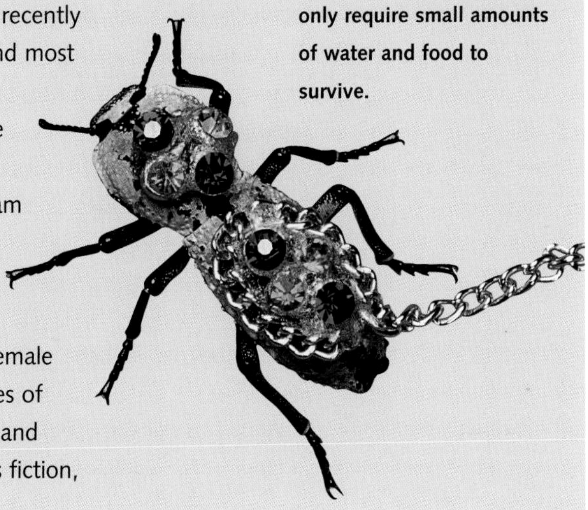

Excavations at Piltdown.

occurred between the early and late Pleistocene, interrupting the last 'Ice Age'. Did Heidelberg Man, as the specimen became known, represent the missing link between humans and apes, or was there, as many experts preferred to believe, an even earlier ancestor? How else, they argued, could scientists account for the existence of eoliths, flints that appeared to have been slightly reshaped to create useful tools, that dated from the early Pleistocene? The discovery of Piltdown Man fitted the bill perfectly. No wonder the follow-up report in *Nature*, published a week after the original announcement, described the skull fragments and mandible as 'the most important discovery of its kind hitherto made in England'.

On 18 December 1912, Dawson and Woodward presented a plaster reconstruction of the skull of the newly named *Eoanthropus dawsoni*, Dawson's Dawn Man, to a packed meeting of the Geological Society. The model was accompanied by a number of flint eoliths discovered in the same location as the fossil remains. The remains were eventually presented to the Museum as a gift that was described in the Trustees report as 'of the highest scientific interest and one of the most important [gifts] ever received by the Department of Geology'.

In a systematic search of the Piltdown gravel pit and associated sites over the next few years, Dawson, Woodward and several others uncovered further human remains, including ten additional fragments of an unusually thick skull, apelike teeth and a further delicate, apelike lower jaw. They also discovered fossil remains from a number of animals, including a Pliocene elephant, a mastodon, a stegodon, a hippopotamus, a beaver, a horse, a red deer and a rhinoceros, and unearthed more eoliths, four Palaeolithic hand axes and, in 1914, a worked piece of bone that became known from its shape as the 'cricket bat'.

Although most experts were convinced that the fossils were genuine, others had grave suspicions about their authenticity. In 1914,

Eoliths from Israel.

William King Gregory wrote, 'It has been suspected by some that geologically [the specimens] are not old at all; that they may even represent a deliberate hoax...', but there was little the experts could do to prove or disprove their doubts.

The only way to calculate the age of the specimens was by studying the rocks in which they were found, and it wasn't until 1953 that experts were able to date the fossil bones themselves, using the latest method of fluorine testing. Led by William Page Oakley, the scientist who developed and pioneered fluorine testing, the team from The Natural History Museum was surprised to discover that the skull was only 600 years old and the jaw was modern. Additional tests showed that the mandible was a fragment from an orang-utan jaw that had been stained to look old, and damaged to remove the parts that would have immediately betrayed its ape origins. There could be no doubt that Piltdown Man was a fraud, and the scientific community was just as enthralled by Oakley's presentation to the Royal Society in 1953 as they had been at the one delivered by Dawson and Woodward forty years earlier.

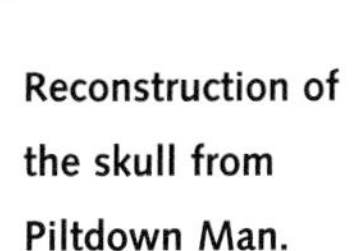

Reconstruction of the skull from Piltdown Man.

There have been many theories about who might have planted the Piltdown fossils. Some believe that Charles Dawson perpetrated the hoax to enhance his scientific reputation. He certainly had the knowledge and was always present, even though he may not have been the person to make the find, whenever a new artefact was discovered. Others point to that master of fiction, Sir Arthur Conan Doyle (1859–1930), who lived nearby, or to the French Jesuit theologian, philosopher and palaeontologist Pierre Teilhard de Chardin (1881–1955), who discovered one of the Piltdown canines while helping with the dig in 1912. But according to Museum scientist Andy Currant and Brian Gardiner of Kings College, London, recent evidence suggests that Martin A.C. Hinton, the leading British Pleistocene expert at that time, and later the Museum's Keeper of Zoology, was the most likely culprit.

Vital new clues emerged in the 1970s, during a clear-out of what was once Hinton's old office at the Museum. A trunk was discovered in the loft that ran above the room, bearing the initials M.A.C.H. It contained correspondence addressed to Hinton, numerous dried-up rodent dissections and, most importantly, a collection of teeth and bones stained the same colour as the Piltdown fossils. Hinton's executor also discovered eight human teeth that had been stained in a variety of different ways, using iron, chromium and manganese. Were these teeth and bones evidence of the experiments used to create the Piltdown fakes?

It was well known that Hinton and Woodward had had an argument in 1910 about money and Hinton's ideas for producing a catalogue of fossil rodents. Was the fraud Hinton's attempt to discredit a man well known for his pomposity and lack of humour? What better way to plot the downfall of an entrenched Darwinian like Woodward than by tricking him into thinking he had found a phoney missing link? Perhaps it was intended as a joke at first, and Hinton underestimated Woodward's lack of expertise in areas outside palaeoichthyology, the study of fossil fish, for which he was, rightly, acknowledged as one of the world's leading experts. It does seem that the fossil finds became increasingly more outlandish as the hoax continued.

The 1914 discovery of the worked piece of bone that looked like a cricket bat must surely have raised a few sceptical eyebrows. And why didn't Dawson and Woodward spot that the animal remains, dug out of the same location, belonged to two distinct faunas from two different eras? The mastodon, stegodon and rhinoceros might have dated from the early Pleistocene, but the red deer, horse, beaver and hippopotamus belonged to a much later fauna. Perhaps Hinton intended the truth to come out far sooner, but neglected to take into account the strength of the need for Woodward, Dawson and many of their contemporaries to believe that they had found the missing link between apes and humans? Like all of the best mysteries, we may never know for sure.

IS IT OR ISN'T IT?

What do *Archaeopteryx* and the duck-billed platypus have in common? They have both been suspected of being part of elaborate hoaxes that have turned out to be completely unfounded. The duck-billed platypus first became known to white settlers in Australia in November 1797, but when a skin was sent back to George Shaw, at the Museum, he was highly sceptical. Part-mammal, part-duck, the fantastic looking creature looked so improbable he thought someone was trying to play a trick. Not to be outwitted, Shaw attempted to prise the bill from the body using a pair of scissors. The bill remained firmly attached, and to this day the impressions left by the scissors can still be seen on the specimen.

In the 1980s, Sir Fred Hoyle, the distinguished astronomer and science-fiction writer, and Chandra Wickramasinghe, professor of mathematical astronomy at the University of Wales in Cardiff, denounced *Archaeopteryx* as a fake. They claimed that someone had created feather impressions by pressing modern bird feathers into a layer of cement or limewash that had been painted around a genuine dinosaur fossil on the surrounding slab of limestone matrix. It was a startling accusation and, like the Piltdown Man hoax, implied that scientists were involved in a plot to fabricate a 'missing link', this time between the dinosaurs and birds.

The Museum issued denials but the media, sensing headline news, seized on the dispute between some of Britain's leading academics and brought it to the attention of the wider public. Unlike the discovery of the Piltdown hoax, however, Hoyle and Wickramasinghes' accusations were completely unfounded, and

Duck-billed platypus by Ferdinand Bauer.

had more journalists taken the time to investigate the facts, and ask why the two astronomers were making the claim, they might have presented the story in a different light.

Hoyle and Wickramasinghe were anxious to discredit *Archaeopteryx* as part of a wider campaign to promote their theory about the evolution of the Universe and life. They were believers of the 'steady state' theory, which asserted that matter was created to fill the gaps left by an expanding Universe. Hoyle and Wickramasinghe also believed that life was only one element within the Universe and that it was present throughout the cosmos. They spent twenty years developing a comprehensive theory that explained the origins and evolution of life on Earth in terms of viruses, supplied by the regular bombardment of comets from outer space, being the main mechanism to 'kick-start' rapid evolutionary novelties.

As proof, the astronomers offered spectrum analysis that revealed the presence of large clouds of organic material in outer space and charts showing that major epidemics, like the influenza attack that killed millions after the First World War, often occurred after a comet impact on Earth. Indeed, they proposed that a comet hitting Earth during the Cretaceous-Tertiary boundary, 65 million years ago, provided an injection of new DNA to produce birds and mammals.

***Archaeopteryx lithographica*, the oldest known bird preserved in 147-million-year-old limestone from Bavaria.**

Hoyle and Wickramasinghe also cited the incomplete fossil record as further evidence in support of their theory. They argued that the view accepted by most biologists, that life had originated and evolved from a single organism, was invalid because of the lack of evidence illustrating a complete branching family tree. Neither could they accept that rates of evolution could have resulted in birds as far back as 150 million years. For Hoyle and Wickramasinghe, a missing link like *Archaeopteryx*, occuring about 82 million years too early in Earth's history, was an unwelcome intrusion that tipped the scale firmly in favour of the traditional evolutionists. *Archaeopteryx* had to go, and what better way than to cry hoax?

It was a poorly conceived idea, not the least because the London *Archaeopteryx* specimen was only one of five specimens, discovered at various intervals between 1861 and 1951, that resided in European institutions at that time. It must have seemed highly unlikely that such an elaborate hoax could have been carried out on separate occasions by three or four generations of different scientists, but the Museum was obliged to refute the claims. The refutation required a painstaking examination of the physical and chemical composition of the fossil and its surrounding matrix to establish whether or not the feather impressions were genuine, a time-consuming task that fell to former Curator of Fossil Reptiles, Alan Charig. The tests showed that there could be no doubting the authenticity of the fossil, and Hoyle and Wickramasinghe lost their credibility as a result of their false accusations. And if further proof were needed, two more specimens of *Archaeopteryx*, complete with feather impressions, have been discovered in the past decade.

They may make interesting stories, but hoaxes, such as Piltdown Man, and contra-frauds, like that relating to *Archaeopteryx*, are a rare side of science that most scientists are keen to forget. Perpetrated by a very small minority, such falsehoods are magnified by the media, and shed a dubious light on a community that otherwise prides itself on its commitment to advancing knowledge through scrupulously fair and honest means.

More importantly, hoaxes and ungrounded accusations waste valuable time. How many precious hours did William Oakley and Alan Charig spend away from busy research programmes, making examinations that, but for the actions of a few individuals, should never have been necessary in the first place?

The Time Detectives and Planet Protectors

'We dwell on a largely unexplored planet.'

Edward O. Wilson, 1992

The Diversity of Life

We share the Earth with millions of other species, many thousands of which we depend on to makes our lives healthier, wealthier and more comfortable. Our economies are built on the trade of natural resources. They provide us individually with the food, clothes, medicine, shelter and a host of other services that we take for granted and yet, despite over four centuries of exploration and discovery, there are still major gaps in our understanding of the great diversity of life on Earth.

To date, biologists have described, named, and classified over 1.7 million species, but many experts believe that this represents only about one-tenth of the world total. As discussed in an earlier chapter in this book, predicting precise numbers is difficult, but on one point there can be no doubt: the number is falling, as many negative effects of human activity, such as global warming, toxic waste, over-harvesting and habitat destruction, continue, to drive species into extinction. In fact, based on calculations like the one made by the great American biologist Edward O. Wilson, who estimates that each year we are losing about 27,000 species in the rainforest, many experts believe we are in the middle of a mass extinction that is

Depressing evidence of the destruction of the rainforest in Peru.

equivalent to the five largest extinction episodes recorded in geological history. They include the Ordovician, Devonian, Permian and Triassic extinctions, and, most recently, the Cretaceous mass extinction that led to the demise of the non-avian dinosaurs.

Killing for meat and hide, and habitat overgrazing by competing livestock, are thought to be the main causes of the extinction of the quagga in South Africa.

Why should we be concerned? With so many species yet to be discovered, many of them may be driven to extinction before biologists have had a chance to study and understand them. We are losing plants, animals and microorganisms that help to maintain the delicate balance of ecosystems and may have important medical applications, or may improve or diversify agricultural productivity.

The giant longhorn timber beetle (*Xixuthrus heros*) measures up to 20 cm in length. It is endemic to the small island of Fiji and is thought to be close to extinction, because of the gradual deforestation of the island and the introduction of predators such as rats.

In 1977, for example, botanist Rafael Guzman discovered *Zea diploperennis*, a new species of wild corn, growing in the cloud forests of Jalisco in Mexico. It is resistant to the seven main types of viral disease that infect cultivated corn. The discovery sparked an intensive-breeding programme, and some of this resistance has now been successfully transferred to domestic varieties. It is hard to put a value on the discovery of *Zea diploperennis*, but in a global market, where corn has an annual turnover of nearly $60 billion dollars, the financial rewards are likely to be huge.

'We make a potentially dangerous mistake when we assume that we must choose between serving humanity or serving the environment. It must be a priority to bring these goals into harmony. They need not and they must not be mutually exclusive.'

Orville Freeman, former US Secretary of Agriculture, 1989

INTERNATIONAL ACTION

Growing concern about the environment and the alarming rate of extinction led to the 1992 Earth Summit in Rio de Janeiro and the first UN Convention on Biological Diversity, where the prime ministers and heads of state of 174 nations recognized the need for continued economic growth, while protecting the integrity of the biosphere. The result was AGENDA 21, an international action plan that listed the objectives set by the Convention on Biological Diversity — the discovery, organization and understanding of the Earth's species — as an urgent priority.

The community of international experts who specialize in systematics, the science dedicated to understanding biodiversity, has been working towards these objectives for years, and has

Peter Melchett

EXECUTIVE DIRECTOR, GREENPEACE, UK

The only blue whale I've ever seen is the huge, magnificent model in The Natural History Museum. Almost no one alive today will see a living blue whale. So many have been killed by whalers that they have almost disappeared from the oceans of the world. The same has happened to almost all the large whales.

I first saw those beautiful model whales in the Museum as a child. Since then, I've been lucky enough to get a job working for Greenpeace, and to campaign to stop any more great whales being killed. Despite the terrible damage whaling has done, two countries, Norway and Japan, are still determined to keep commercial whaling going. The United Nations has said that whaling should be controlled by the International Whaling Commission (IWC). After years of campaigning by Greenpeace and others, the IWC has imposed a moratorium on commercial whaling. Japan and Norway ignore this. They flout the moratorium, carrying on killing the one whale species that hasn't yet been decimated by whaling, the minke whale.

Most minke whales live inside a whale sanctuary in the waters around Antarctica, and incredibly it is here

Hauling in a harpooned minke whale.

that Japanese whalers do much of their killing. Minke whales also live around Scotland, in the North Sea and the Atlantic. It was there I saw my first live whale. I visited the remote and beautiful island of St Kilda,100 miles out in the Atlantic from the Scottish coast, to look at the threat that oil developments pose to its incredible seabird colonies, and to whales and dolphins. As we plunged through heavy seas, a minke whale surfaced in a shower of spray and was gone. A magical moment.

Greenpeace and the Whale and Dolphin Conservation Society carried out the first dedicated, scientific survey of whales and dolphins in those deep and stormy waters. Our survey team saw nearly 1000 whales and dolphins in just over two weeks at sea! I suspect very few visitors to the Museum realize, as they look at those model whales, just how many live close to our shores, including fin, sei and minke whales. We even have the giants of the oceans, blue whales, living in those same waters. We know this because their songs were heard by US Navy listening devices, placed at the bottom of the Atlantic to listen out for Russian submarines.

Greenpeace will carry on campaigning until Norway and Japan stop commercial whaling, and minke whales like the one I saw off St Kilda are safe from that threat. But whales face other dangers. Toxic chemicals, which we release into the environment, build up in their bodies. The damage to the ozone layer threatens the krill many whales feed on. And climate change threatens the stability of ocean currents and food webs which all marine life, including whales, depends on.

identified three priorities for systematic biology research if we are to achieve them on a global scale:

- 'To discover, describe and inventory global diversity.'
- 'To analyse and synthesize the information derived from this global discovery effort into a predictive classification system that reflects the history of life.'
- 'To organize and communicate the information derived from this global programme in an efficiently retrievable form that best meets the needs of science and society.'

The UK has an enormous contribution to make towards this international effort because of the scientific and historical importance of its collections, and its long tradition of research in systematics. British experts have set themselves three objectives: to use these invaluable resources, to support international research programmes, and to promote the conservation and sustainable use of biological resources within the UK. This chapter will look at how collections-based research is helping scientists to contribute to these aims, in addition to solving a range of problems in areas such as human health, veterinary medicine, forensic science and agriculture.

The biological inventory of species in the UK is far from complete. New species are always being discovered. Around 100 new lichen records have been added since 1992, including a significant number of new species.

The Discovery, Description and Inventory of Global Diversity

'Every country has three forms of wealth: material, cultural and biological. The first two we understand well because they are the substance of our everyday lives. The essence of the biodiversity problem is that biological wealth is taken much less seriously. This is a major strategic error, one that will be increasingly regretted as time passes.'

Edward O. Wilson, 1992

Previous chapters have shown how contemporary taxonomists have continued to discover, describe and inventory biodiversity throughout the 20th century, so why are they such a long way away from completing a global biological inventory? Experts have identified a number of obstacles, or 'taxonomic impediments', that have prevented them from filling in the gaps in their understanding of global biodiversity. One reason is that some of the most biologically rich nations lack the necessary funds to invest in a national programme of systematic research. While 94% of the world's systematists, and the majority of its collections and research centres, are based in the wealthier, more developed countries, estimates show that 80% of the Earth's terrestrial diversity is to be found within the tropics, a region that includes some of the most economically impoverished countries on Earth.

There is also an imbalance in our knowledge of species, distribution and evolutionary relationships between different groups of organisms. We may have developed considerable knowledge of groups such as birds and mammals, but little is known about some

Las Cuevas Research Station, a joint initiative between The Natural History Museum and the Government of Belize, to investigate the biological diversity of these tropical forests.

of the other groups, such as the prokaryotes, which are the most species-rich, and which potentially may be the most economically or ecologically important. This is partly due to the fact that there is a shortage of systematists with the specialist knowledge to work on them. A 1994 survey of expertise in systematic biology in the UK, for example, showed that of a total of 421 professional systematists, 168 were invertebrate specialists, while only 17 were prokaryote specialists.

Understanding these taxonomic impediments has been the first step towards developing a strategy to overcome them, and has led to some exciting new research projects for the Museum. In 1989, for example, the government of Costa Rica, one of the most biologically diverse countries on Earth, set up an ambitious programme to make a national biological inventory and conserve its natural resources. The project was set up through the Costa Rican National Biodiversity Institute, INBio (Instituto Nacional de Biodiversidad), using the advice and expertise of scientists working at the Museum.

Participants in a parataxonomy course at INBio, Costa Rica.

The Museum's Department of Entomology has provided a number of intensive training courses for INBio's parataxonomists, the field workers responsible for collecting, and external funding has been raised for other Museum staff to train several INBio biologists in London and supervise a number of Costa Rican MSc students. Museum experts are also collaborating with Costa Rican scientists on various research projects and the national biological inventory. This includes an extensive Malaise trap survey of the hymenopterous fauna of Costa Rica. The Hymenoptera, including bees, wasps and ants, are an insect group of major ecological and economic importance as plant pollinators or predators and parasitoids (species in which the larval stages are parasites of the host's larvae) of insect herbivores. This survey is a major undertaking, providing the basis for study for more than 20 international systematists working on an introduction to the hymenopterous fauna of Costa Rica. It is the first time such a major work has ever been produced for a tropical country.

UNCOVERING THE PATTERNS OF EVOLUTIONARY HISTORY

The discovery and identification of species enable experts to list the species that share our planet. The next objective is to describe the characteristics of each species, and make comparisons, using these and a wide range of data collected from many other fields —

The common yew, *Taxus baccata*.

including anatomy, biogeography, developmental biology, ecology, ethology (the study of animal behaviour), geology, palaeontology, physiology and, most recently, genetics and molecular biology. These comparisons allow experts to establish the most likely evolutionary (phylogenetic) relationships among species and draw up a natural classification system that allows them to make predictions about species and their properties, which may be of particular value to science or society at large.

The Pacific yew, *Taxus brevifolia*, for example, produces a natural product from its bark called taxol, which has been found to be an effective drug against ovarian and breast cancer. It takes the bark of three trees to supply enough taxol to treat one patient, however, and because the trees are killed in the process, experts started to look around for an alternative source. A random search for plants with similar properties might eventually have pinpointed a suitable candidate, but understanding the evolutionary relationships of the Pacific yew enabled American researchers to target its closest relatives immediately. They were delighted to discover that the European yew, *Taxus baccata*, also produces taxol and, better still, that only a small number of leaves were necessary to produce the same yield without destroying the tree. Similarly, when the Moreton Bay chestnut, *Castanospermum australe*, was found to produce a compound with promising anti-HIV properties, researchers at Kew Gardens, London, were able to save valuable time and research money by directing researchers towards closely related species that were most likely to produce compounds with similar pharmaceutical properties.

The following case studies show how collections-based research is helping scientists contribute to our understanding of evolutionary history and, in the case of research on ancient hydrothermal vent communities, how this work is helping us to understand global change over the past five hundred million years. Experts believe that understanding global change in the past will help them to predict change in the future, and so help conservationists and policy-makers set priorities for the sustainable use of natural resources.

DNA — A MESSENGER FROM THE PAST

Rapid advances in the understanding of genetics over the last 20 years has provided systematists with another important means of testing the evolutionary relationships among species. Extracting DNA from samples of living organisms has become a routine process, but it has proved extremely difficult to gather samples from organisms that have been dead for a long period of time, unless they have been preserved under special conditions. Techniques are improving all the time, however, and some scientists have been successful in extracting DNA from the remains of dead organisms,

Jana Bennett

SENIOR VICE-PRESIDENT OF DISCOVERY COMMUNICATIONS INC.

For many visitors, the age of the dinosaurs is what The Natural History Museum promises as you enter the Central Hall, and it is hard to meet the *Diplodocus* peering down at you without marvelling. However, the Museum offers many less obvious wonders, and to me one of the most fascinating has to be the dodo. I don't choose it for the quality of the example. The dodo is special because humans were the agents of its disappearance. While we may wonder at the age of the sauropods and the past dominance of the dinosaurs, the dodo has the distinction of being the first recorded creature to be made extinct directly by humans. Yet, the curious thing is that the dodo was considered valuable alive; dodos were being captured and exhibited live in Europe just as the species was being allowed to die out in its home on the Indian Ocean island of Mauritius.

***The Dodo and Kindred Birds*, Roelandt Savery, 1625.**

The dodo, a flightless 11 kg pigeon, was first reported by Dutch sailors when they landed on Mauritius in 1598. There they found the strange birds and their large eggs, with few defences and few natural predators. The sailors, and the pigs and rats introduced to the islands, all fed off the dodos' eggs, and contributed to the rapid extinction of the birds on mainland Mauritius, probably as early as the 1640s, just 40–50 years after discovery. A report from some ship-wrecked seamen in 1660 suggested they still survived on islets close to shore. This was the last definite account.

Dodos were a curiosity, so they were collected by early travellers to present to their country. Generally, they were eaten, but a bird that could survive the hardships of an ocean crossing and reach European shores alive would have had great potential for ornamental purposes in collections. Sadly, no attempt was made to breed them or get any numbers together. Most specimens collected during the 17th century are now lost — not surprisingly, as many of the early collections were private. No one at the time could have realized that no more specimens could be procured, and it was often assumed that a replacement for a disintegrating specimen would easily become available.

The dodo symbolizes the story of exploration. Its fate is part of the human voyage, from an age of exploration and colonization, when collecting took the form of an uncontrolled pre-scientific harvest, to today's fieldwork, which combines conservation with study in the field. Humans started to bring about its extinction by killing the birds directly and by introducing new predators, and finished it off through their passion for collecting. The dodo's story is poignant, mysterious and relevant. The exhibit links us directly, as visitors to the Museum, both to extinctions in the past and to the issue of the future diversity of species. That is why I linger in front of the dodo.

Jana Bennett

including some that have become recently extinct. What have they been able to learn from this new source of information? Jeremy Austin explains how studying the DNA of extinct giant tortoises is helping to shed light on how they evolved:

'Four hundred years ago humans set foot for the first time on three isolated volcanic islands in the Indian Ocean — Mauritius, Réunion and Rodrigues — that lie 750–1700 kilometres away from Madagascar, the next nearest island. The islands supported a rich variety of flora and fauna, including unique giant land tortoises with shells that were only a few millimetres thick, much thinner than the shells of giant tortoises found on Madagascar and the Seychelles. The shells of tortoises form an integral part of their skeleton, and it is thought that the island tortoises had somehow arrived on the islands from elsewhere and subsequently evolved to reduce their shells to the bare minimum necessary to support their body. They had also developed much bigger openings in the shells for their head, arms and legs. Evidence suggests that before the arrival of humans, who subsequently hunted them to extinction, there were no predators of tortoises on all three islands, and therefore no advantage for the bigger, heavier protective shells found in other land tortoises.

Sadly, only limited evidence of the giant land tortoises that once inhabited Mauritius,

Giant tortoise, 1990, a chalk drawing by Bryan Kneale.

Réunion and Rodrigues exists today. All that remains are a few bones and some fragments of shell, not enough to use in conventional morphological techniques to answer all the questions that my colleague, Nick Arnold, and myself have about the evolutionary relationship that exists between the different island tortoises.

We knew that there were two forms of giant land tortoises on Mauritius, Réunion and Rodrigues. One was a grazer with a dome-shaped shell, and the other, with a saddle-backed shaped shell, was able to reach up and browse on leaves, but we weren't sure whether or not they represented separate species or how they came to colonize all three islands. We believe that the island tortoises had originated from giant land tortoises that had drifted across the sea from Madagascar, Africa or the Seychelles. There have been a number of independent observations of giant tortoises feeding on seaweed at low tide and then being washed out to sea at high tide. With their fat reserves, and the arrangement of their lungs acting like buoyancy bags at the top of their shells, it seemed reasonable to assume that some tortoises had survived the journey across the sea to colonize the three volcanic islands. But the question was: had this colonization taken place once in Mauritius and did speciation (the evolution of new species) take place a single time to create the domed-backed and saddle-backed species? Did they then get washed across to Rodrigues and Réunion? Or did the tortoises spread from Mauritius to Rodrigues and Réunion first, and speciation occur three times independently on each island?

If we could extract DNA from the bone specimens from each island, we could compare the samples, and create a family tree to show the relationships within the group, and answer the questions that had been eluding us. So far, our initial results have been promising. We have been able to extract DNA fragments from a number of bones from all three islands, and the DNA sequences are providing the first glimpses of relationships among these island tortoises. Definite answers to the riddle of the evolution of these island tortoises will come when we have DNA sequences from all the species.'

WERE THE NEANDERTHALS OUR ANCESTORS?

Leading anthropologist Chris Stringer looks at the latest evidence based on research he conducted with his team of colleagues from The Natural History Museum:

'The Museum has a long association with research on the prehistoric people known as Neanderthals. The Palaeontology Department curates some of the most important Neanderthal fossils, and these are a continuing focus of research by visiting scientists and our own staff. And over the last five years, Museum staff have been excavating at Neanderthal sites in Gibraltar. The Neanderthals got their name from a fossil skeleton found in the Neander Valley (Neanderthal) in Germany, which was first scientifically described in 1857, two years before Darwin's *On the Origin of Species* was published. This not only marked the real beginning of the field of palaeoanthropology, but also ignited an intense debate about the relationship of these extinct people to us, and in particular to those of us who are of European descent.

An adult female Neanderthal skull from Forbes Quarry, Gibralta.

Some scientists of the time believed that the distinctive features of the first Neanderthal skeleton were due to disease, but as other similar finds were made across Europe, it became increasingly clear that these fossils represented an ancient population of humans who inhabited Europe during the last Ice Age. Even before the German find, a similar skull had been discovered in Gibraltar, but it was only brought to England for study in 1863, and its importance was not recognized until later. This skull, from Forbes' Quarry, is now part of our collections, along with a Neanderthal child's skull found at the neighbouring site of Devil's Tower in 1926.

We now know that the Neanderthals originated in Europe over 200,000 years ago, and disappeared about 30,000 years ago. The situation is complicated by the additional presence in Europe, from about 35,000 years ago, of much more modern-looking people known as the Cro-Magnons, who seem ultimately to have originated in Africa. However, experts remain fiercely divided over the question of whether the Neanderthals were replaced by the invading Cro-Magnons, perhaps aggressively or through competition for the same resources, or whether they mixed and mated with each other — in which case, modern Europeans might be partly descended from the Neanderthals.

JURASSIC PARK — FACT OR FICTION?

The film Jurassic Park is based on the notion that ancient dinosaur DNA can be extracted from fossilized mosquitoes preserved in amber. It assumes that after feeding on dinosaurs, these prehistoric mosquitoes became trapped in certain tree resins and preserved, retaining traces of dinosaur blood, and therefore dinosaur DNA, in their guts. The resin solidified and eventually, after millions of years, hardened to form beads of amber, minute time capsules containing perfectly preserved dinosaur DNA, that Jurassic Park's scientists use as genetic 'blueprints' to clone living dinosaurs. So how close is this fictitious story to the truth? Expert in ancient DNA, Jeremy Austin gives us his views:

Q There have been recent reports in the media about real-life attempts to extract ancient DNA from insects preserved in amber. How easy is it to isolate DNA from this source and has anyone been successful?

DNA is extremely fragile, degrades in water, and tends to fall apart and lose its 'signature' very easily. Research has shown that the DNA of dead organisms begins to fragment very rapidly unless it has been preserved under unique conditions. Preservation in amber seemed to offer a reasonable option, and there were several reports, including the one in 1991 that claimed that DNA fragments had been recovered from insects — in this case, Dominican bees that had died between 25 and 125 million years ago. These reports caused considerable excitement but, despite intensive efforts, no other researchers, including the team at The Natural History Museum, have been able to repeat and verify these results. As a result of these findings, most scientists now agree that DNA doesn't survive in fossilized insects in amber.

Q Do you think it is likely that scientists will ever be able to extract dinosaur DNA from fossil mosquitoes in amber?

We have no record of fossilized mosquitoes that are as old as the dinosaurs, and even if specimens were found, I think our lack of success in isolating ancient insect DNA from specimens preserved in amber make the chances of isolating dinosaur DNA from the same source virtually non-existent.

Q Has anyone been able to isolate dinosaur DNA from other sources, such as fossilized bone?

Fossilized bone consists of minerals that have replaced the original organic material and this, coupled with the fragility of DNA, makes it extremely unlikely that even fragments of dinosaur DNA will have survived to modern times. There have been recent preliminary

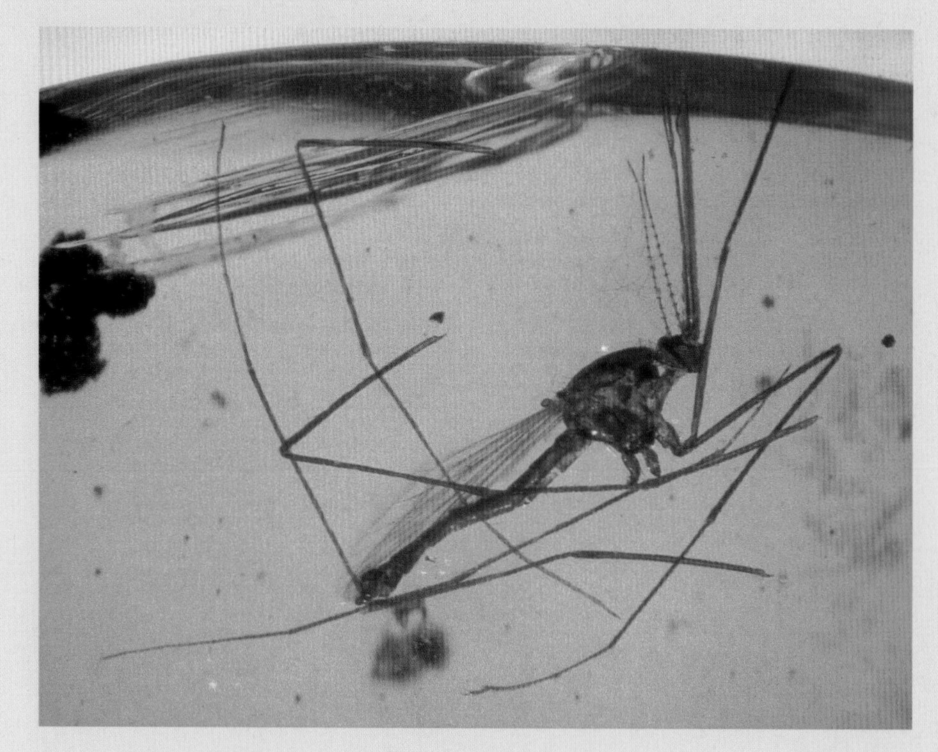

Mosquito in Dominican amber.

reports claiming successful isolation of ancient DNA from dinosaur bone, but no scientist has been able to prove conclusively that they are indeed fragments of dinosaur DNA. One of the difficulties in verifying the identity of these fragments is that DNA is a constituent of all living organisms, and the highly sensitive experimental procedures that are used to isolate fragments of ancient DNA are susceptible to contamination from contemporary sources. Scientists can inadvertently isolate DNA from organisms that live in the soil surrounding the fossil specimen, for example, or from their own skin.

Q Do you think it likely that scientists will ever be able to extract enough DNA from fossilized remains to reconstruct the complete DNA sequence for a dinosaur?
I think it would be an impossible task. Even if we could successfully isolate fragments of dinosaur DNA, mapping the correct DNA sequence for a complete dinosaur genome would be like trying to predict the contents, and order, of a complete library of information from the facts contained in just one or two pages of a single book.

To give you an idea of the complexity of the problem, I have been analysing DNA from specimens of lizards that became extinct about 300 years ago. The largest single piece of DNA sequence that I have been able to obtain so far, after six months of effort, is 50 base pairs. These minuscule snippets of information tell me that I am looking at a lizard from a particular group, but it is a quantum leap away from enabling me to predict the thousands or millions of base pair sequences for a single feature of the animal — let alone its complete genome.

Q What do you think about Dr David L. Stearn's recent prediction that, in years to come, DNA mapping techniques will enable scientists to map a complete dinosaur DNA sequence by working backwards from the DNA sequences that they will eventually unravel for birds, reptiles and mammals?
Scientists have successfully mapped the complete DNA sequence for two bacteria (*Haemophilus influenzae* and *Methanococcus jannaschii*), a yeast (*Saccharomyces cerevisiae*), and a nematode worm (*Caenorhabditis elegans*), and there are research projects underway to try and map the complete DNA sequence for humans and the fruit fly *Drosophila*, but they will take years to complete. The Human Genome Project, for example, has research teams collaborating around the world to map the three billion base pairs that constitute the human DNA sequence, but even with this collaborative effort it will have taken close to 15 years to complete.

Although it may be possible to map the DNA sequences for birds, reptiles and mammals in theory, I'm not convinced that this is going to happen in the near future, because of the enormous cost and effort involved. And even if complete genome sequences were available for living mammals, birds and reptiles, predicting the sequence of a dinosaur genome with any degree of accuracy would be an impossibly complex task. Think of the dinosaur sequence that we are trying to predict as a hypothetical jigsaw puzzle made up of more than one billion pieces, and each piece, instead of being a flat piece of card, is a cube with a different fragment of the overall picture on each side. So to reconstruct this jigsaw puzzle you not only need to position all the pieces in the correct place, but you also have to have the correct face showing too. That's how impossible the problem is!

Q Stearn also predicts that about 200 years from now, having completed the dinosaur DNA sequence, scientists will be in the position to build the first dinosaur made by humans. Could this be the basis upon which the fictitious Jurassic Park finally becomes a reality?
This is all highly speculative. Even if scientists were able to reconstruct successfully the DNA sequence for a dinosaur genome, how would they 'clone' this animal from its genetic blueprint? I think the sheer complexity of first sequencing the genome and then finding a way of using this information to build a living organism makes it an insurmountable task, and I personally remain convinced that dinosaurs will safely remain fascinating creatures that belong to the past.

New finds and new techniques of investigation have increased our knowledge of this critical time period in human evolution, but some research has also raised new questions about what really happened to the Neanderthals. Museum staff are closely involved in research on the Neanderthals, and they have recently returned to Gibraltar to find out more about the Neanderthals who once lived there.

From fossil remains, we know that the physique of the Neanderthals was muscular and thickset, with a short, wide body reflecting adaptation to the cold of the Ice Age, and a skeleton built for strength and endurance. Their brains were as large as ours, but enclosed in a longer, lower skull, with a strong brow-ridge over the eyes. The face was large and long, and was dominated by an enormous nose, perhaps part of an adaptation for breathing in relatively cold, dry air. The front teeth were relatively large and often heavily worn, and the lower jaw lacked a prominent chin. Recent research on our Neanderthal fossils, using computerized three-dimensional x-ray imaging, has revealed further details of their anatomy, even down to the fact that their inner ear structure was slightly different from ours (the reasons for this are, as yet, unclear).

The Neanderthals were mobile hunter-gatherers who lived off the land, collecting plant resources and hunting or scavenging large and small game. They made effective, but relatively simple, tools and, although they had the very human behaviour of burying their dead, many archaeologists have questioned whether they had other modern human attributes, such as complex language or art. In seeming contrast, the Cro-Magnons, also hunter-gatherers, certainly produced sophisticated tools and art, including the spectacular paintings of the Lascaux and Chauvet caves, and most palaeoanthropologists consider that they were very similar to us in both anatomy and behaviour.

Maurice Wilson's reconstruction of Neanderthals outside Gorham's Cave in Gibraltar. The painting is based on the remains found within various caves in the area. One of the hunted birds, the great auk, is now extinct.

One of the most remarkable developments of recent years was the recovery, in 1997, of fragments of ancient DNA from the original Neanderthal skeleton, providing our first glimpse of the genetic make-up of an extinct form of human. Comparisons of this DNA with that of modern humans and chimpanzees imply

LEFT: **Excavation of a Neanderthal fireplace at Vanguard Cave, Gibraltar.**

RIGHT: **Gorham's Cave, Gibraltar, as it appears from the sea today.**

that the evolutionary divergence between modern humans and Neanderthals might have begun about half a million years ago, although this estimate is, of course, based on data from only a fraction of the whole DNA of one Neanderthal individual.

Nevertheless, analyses also failed to support any particularly close relationship between modern Europeans and the Neanderthal, since recent European DNA was as distinct from it as that of Africans, Australians or Asians. Further DNA analyses are in progress on other Neanderthal material, including some from the Museum.

Late in 1998, another significant discovery was made at a site called Lagar Velho in Portugal. A fossil skeleton of a child, dated about 25,000 years old, was excavated, and it has been claimed that this shows a mixture of Neanderthal and modern features. It is described as combining a modern-looking lower jaw and teeth with the robust, cold-adapted body shape characteristic of Neanderthals. However, it is to soon to evaluate the claim that it is a hybrid in detail. It is possible, for example, that the young age of the child has affected the degree of expression of either modern or Neanderthal features. It is also not yet known whether it shows Neanderthal or modern features in significant regions, such as the upper front teeth and inner ear. Alternatively, since we know that Portugal was significantly colder during this part of the Ice Age, could it represent the first evidence of modern humans adapting to the cold of the last Ice Age, thus paralleling the Neanderthals in some respects? These questions can be answered only by further research.

Neanderthal artefacts from Gorham's Cave, Gilbraltar.

If the Lagar Velho skeleton really is that of a hybrid, it would conclusively demonstrate that interbreeding did take place between the last Neanderthals and the Cro-Magnons, but it cannot tell us how common such matings were, whether the hybrids were fertile, whether their genes penetrated into Cro-Magnon populations, and what was the eventual fate of such genes today, after some 1500 generations. And the existence of hybrids would not even disprove the idea of some anthropologists, including me, that based on their distinctive anatomy, the

Neanderthals were probably a different species from us. Such a species would have been very closely related to us genetically, and species of mammal do hybridize — for example, lions with tigers (in captivity), and dogs with wolves. The real question would be whether the species merged, and we have no evidence of that for the Neanderthals and Cro-Magnons, since their core populations seem to have remained quite distinct. Fossil anatomy and recent DNA studies still support the view that Neanderthals were not our ancestors. Neanderthal DNA recovered so far is quite unlike that of people alive today, and extensive genetic sampling of present-day Europeans has so far found nothing resembling it.

However, recent research on Neanderthal living sites certainly requires us to upgrade our view of their capabilities. At the site in Arcy-sur-Cure, central France, for example, some of the last Neanderthals appear to have made jewellery of mammoth ivory and animal teeth, and a fierce debate has ensued about whether they did this independently, or only under the influence of neighbouring Cro-Magnons. And, in contrast to the prevailing view that Neanderthals did not use marine resources, our excavations in Gibraltar show that 50,000 years ago at Vanguard Cave, their varied diet included baked mussels, which they had harvested at least a mile away. Such behaviour seems very familiar, and shows us that even if the Neanderthals were not our ancestors, ours was not the only path that led to humanity.'

Global Change — The Past is the Key to the Future

Over the past 20 years, the term 'global change' has become synonymous with the word 'disaster' in a world increasingly concerned about the ways in which human activities are upsetting the delicate balance of nature. But as the last two centuries of study into the history of our planet has shown, neither the Earth nor its inhabitants have ever remained static. The fossil record tells a story of long periods of relative stability punctuated by successive mass extinctions — sudden, fundamental changes to the Earth's flora and fauna. And geological surveys reveal evidence of continuous environmental change across the scale from plate tectonic movements and volcanism to fluctuations in global temperature and patterns of ocean circulation. About 145 million years ago, for example, the Earth's climate was generally warm and large areas of its continents were covered by sea. But today we live in a cooler period, with a significant percentage of the ocean 'locked up' within the polar ice caps, keeping sea levels lower than for most of geological history.

Studying patterns of change in the past tells us that the Earth is heading for another glacial period, but what concerns the experts is that we might be greatly accelerating the rate of current global warming. Increased greenhouse gas emissions may increase global temperatures

Is large-scale farming like this in Montana, USA, affecting climatic belts?

Petrochemical plants such as this in West Lothian, Scotland, could be contributing to increased greenhouse gas emissions.

and produce a rapid rise in sea level, the disruption of coastal ecosystems and shifts in the Earth's climatic belts. And although we know that terrestrial organisms have survived rapid climate changes by following migrating climatic belts in the past, the extent to which we have damaged ecosystems and developed agricultural areas may have cut off the escape routes that will enable them to migrate in the future. Making predictions about the effects of our actions today on the planet in the future is extremely difficult, but experts believe we can only improve them by understanding the ways in which environmental changes have affected the Earth's flora and fauna in the past. The following example describes how expert Crispin Little's research on the fossil animals associated with ancient hydrothermal vents is beginning to help us understand how these communities have responded to global change over five hundred million years.

ANCIENT HYDROTHERMAL VENT COMMUNITIES

The Ural Mountains have been mined for their rich mineral deposits for over 100 years, but it wasn't until 1947 that Russian scientists first described what appeared to be unusual fossil animals preserved within metal ore samples found in one particular site — the 340-million-year-old Sibay deposit that dated from the Devonian period. There were many ideas about how these zinc, copper and iron sulphide deposits (known as massive sulphide deposits) had formed. One hypothesis suggested that they had formed on the ocean floor, another that they were produced during the formation of the Ural Mountains, and the discovery of the fossils in the Sibay deposit only added to this speculation.

In 1977, the discovery of modern hydrothermal vents in the Pacific Ocean enabled Russian scientists to reinterpret the evidence and construct a more convincing explanation. As experts started to investigate the mineralogy of samples taken from the deposits that precipitated from modern hydrothermal vent fluid in the Pacific, Valeriy Maslennikov and Viktor Zaykov, the Russian scientists studying the mineralogy of different deposits in the Urals, noted important similarities which suggested that the Ural deposits may have formed from fluid produced by ancient hydrothermal vents. Geological surveys provided additional evidence to support the hypothesis. These showed that during the period of geological time in which the various deposits were formed, the area of land occupied by the Ural Mountains was once covered by ocean. The ocean filled a gap between the East European continental plate and the Siberian continental plate, but gradually disappeared over millions of years as they converged and the edge of the overriding plate crumpled and uplifted to form the mountain chain.

Even more convincing was the discovery in the 1980s of a second source of fossils that were much older, yet better preserved, than those found in 1947. A detailed study of these 430-million-year-old fossil animals, found in

On site at Yaman Kasy mine in the Southern Urals.

the Yaman Kasy deposit that dated from the Silurian, revealed close similarities with some of the newly described species found within modern hydrothermal vent communities.

A description of the fossil fauna, including molluscs, brachiopods, and large tube worms was published in Russian in the early 1990s, but because of the language barrier it failed to attract the attention of the English-speaking scientific community until several years later. Palaeontologist Crispin Little and mineralogist Richard Herrington finally made contact with Valeriy Maslennikov and Viktor Zaykov in 1995 to establish a program of collaborative research. Crispin Little takes up the story:

430-million-year-old fossil tube worms from the Yaman Kasy mine, the site of an ancient hydrothermal vent.

'The discovery of these ancient hydrothermal vents was extremely exciting. Like many other experts, we knew that if the Urals marked the site of ancient hydrothermal vents, there must be other land-based sites, produced under similar conditions, all over the world. And this would provide a rare opportunity to study patterns of change within the fauna associated with these sites, together with environmental changes that have taken place over hundreds of millions of years. My tasks were to first identify, describe and classify the fossil animals found at each site and then reconstruct the vent communities, using the taxonomic information to build up a picture of the interactions amongst the various species. Meanwhile, Richard studied the mineralogy of the various ancient deposits to work out the conditions in which these animals lived. His investigations revealed fragments of ancient black-smoker chimneys, providing yet another source of evidence to support the hypothesis that these massive sulphide deposits had been produced from vents on the ancient sea floor. Richard showed that the fluid from these ancient hydrothermal vents reached temperatures of 370°C, equivalent to the highest temperatures found in modern hydrothermal vents.

After Richard and I had spent three years working with our colleagues on the ancient hydrothermal vents in the Urals, I began to look for evidence of other sites around the world. Most recently, this has included an expedition in search of fossilized vent fauna in the Troödos Mountains in Cyprus, the Mediterranean island famed throughout history for its rich supplies of copper. Although the Troödos Mountains had been mined for pyrite and copper for over 2000 years, there were no convincing records of any fossils found within these ores until 1990, when Michel Morisseau, a geologist based on the island, reported fossil finds in one of the deposits. Michel's discovery of 91-million-year-old tubeworms and gastropods that dated from the Late Cretaceous period seemed far more promising. And I believed that if Michel had found fossils at one site there was a good chance that there would be fossils at other sites in the Troödos Mountains, too.

I flew to Cyprus in 1997, full of optimism that Michel's interpretation was correct, even though I couldn't be certain that he had found ancient vent fauna until I examined the fossils first-hand. The geology of the Troödos Mountains certainly seemed to support the possibility that they were formed, like the Urals, from oceanic crust at the boundary of two converging tectonic plates, the European plate and the North African plate. And although I knew that thousands of people must have examined mineral deposits from the Troödos over the years, I persuaded myself that to the untrained eye the fossil animals that we were looking for could easily have escaped detection.

Any lingering doubts instantly evaporated the moment I laid eyes on the fossil animals that Michel had discovered. There could be no doubt that they were members of an ancient vent community, and I felt confident that we would soon find other fossils. Despite this optimism, I never imagined for one moment that I would find one within minutes of stepping into the Kambia Mine, the first location that I visited. There in my hand, in the third piece of pyrite, randomly selected from the hundreds of samples that littered the spoil pit at the side of the open-cast mine, was a perfectly preserved fossil tubeworm.

We went on to find many more fossils during my two-week visit, and are continuing to build up a detailed picture of the vent communities that flourished in this region 91 million years ago. This research is still in progress, but early comparisons with the fossil fauna and contemporary vent communities from other sites around the world are starting to produce some interesting results.

One unusually fossil-rich vent site, for example, was discovered in a 187 million-year-old deposit in the San Rafael Mountains, near Santa Barbara in California, that dates from the Jurassic period. Studying it indicates that although the conditions in the vent have remained fairly constant over hundreds of million of years, the communities of animals that live around them have been far more dynamic. The San Rafael fossil vent fauna has many animals, including vestimentiferan tube worms and gastropods, in common with modern vent communities. But it also includes

View towards Adelphi Ridge, Troödos Mountains, Cyprus.

Cast of a gastropod shell in sulphide ore from Troödos, Cyprus.

rhynchonellid brachiopods, an important group of animals that have been found in other ancient deposits like Yaman Kasy, but aren't known from modern vent sites.

It also appears that several groups of animals found in modern vent communities — crustaceans, for example — weren't present in any of the ancient vent communities. This difference suggests that there have been movements of important groups of animals in and out of vent communities over long periods of time, and in many ways this has been a surprising discovery. The mineralogical evidence suggests that vent environments have remained relatively constant through time, and before experts began investigating them, they thought the same would be true of hydro-thermal vent fauna. We don't know why changes within hydrothermal vent communities have occurred, but they do seem to reflect the changes in biota that we see throughout the Earth's history, and this suggests that they have been influenced by factors beyond conditions within the local environment.'

Beyond Planet Earth

While scientists struggle to learn about the mysteries of life from their explorations of our own planet, they also look into the Solar System to understand more about the origins of the Earth and that of the worlds beyond. How do the experts explore these distant planets and what can we learn from them? Martian-meteorite specialist Monica Grady explains:

'Our local star, the Sun, was formed out of a nebula, a rotating and turbulent cloud of gas and dust. As the cloud rotated faster, it collapsed and flattened into a disk with the Sun at its centre. Within the disk, dust grains joined together to form bigger and bigger bodies, eventually, some 4560 million years ago, producing the planets and their satellites, plus the host of other bodies that make up the Solar System.

The four inner planets (Mercury, Venus, Earth and Mars) are made mainly from rock. Then follow the giant planets Jupiter and Saturn (composed mainly of gas) and the outer

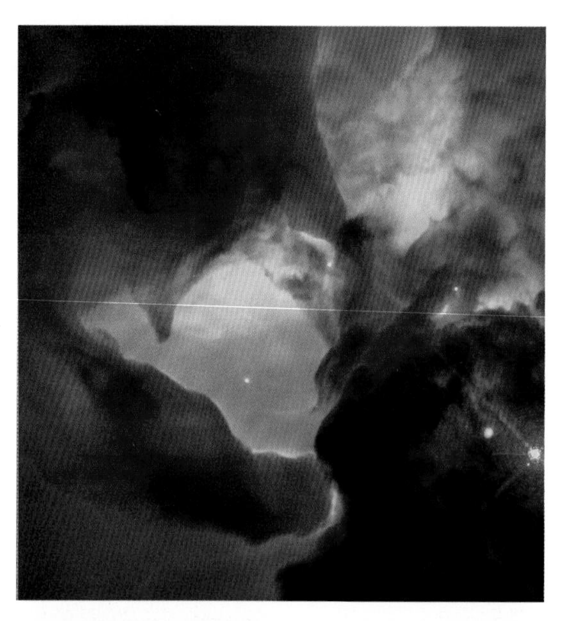

Lagoon nebula.

The giant planet Jupiter.

planets Uranus and Neptune (gas plus ice). The outermost planet, Pluto, and its satellite, Charon, are small compact icy bodies that have an affinity with a disk-like array of similar objects known as the Kuiper Belt.

The icy bodies that produced comets are now thought to inhabit a spherical region of space, called the Oort cloud, extending to about 50,000 AU (1 Au is the mean Earth–Sun distance, approximately 150 million km). Like the planets, comets orbit the Sun: for example, Halley's comet takes 76 years to complete one orbit. Comets have been described as 'dirty snowballs', a mixture of ice and dust that has never completely melted. Each time a comet approaches the Sun, part of its ice melts, and streams away from the central portion, or nucleus, of the comet, carrying with it some of the dust. This expanding cloud of gas and dust gives rise to a comet's characteristic head and tail. Each time the comet draws near to the Sun, more of the ice and dust is lost, and the dust eventually becomes spread out along the entire orbit of the comet.

Between Mars and Jupiter, at a distance of approximately 450–700 million km at the hiatus between rock and gas in the Solar System, lies the asteroid belt, the source of most meteorites. There are several thousand asteroids, the largest of which, named Ceres, is about 914 km across (for comparison, the Earth's diameter is about 13,000 km, and the Moon's about 3500 km). Asteroids are rocky, metallic or carbonaceous bodies. They are material remaining after the planets formed (Jupiter's gravitational pull prevented the bodies from joining together to form a single planet). Occasionally, influenced by Jupiter, the orbit of an asteroid is altered, such that it might collide with another, and break up. Fragments of disrupted asteroids fall to Earth as meteorites, natural objects that date from the formation of the Solar System, and comprise the only source of primordial material available for laboratory-based study. Although the Earth, along with the other planets, was also formed 4560 million years ago, none of the original material remains: it has been removed by bombardment or otherwise eroded, or recycled through geological activity (such as plate tectonics and volcanism). It is only by studying meteorites that scientists can learn about the processes and materials that shaped the Solar System and our planet.

Comet Hale Bopp visible during sunset.

Meteorites are divided into three main types (stone, iron and stony-iron), reflecting their composition. Different types of meteorite provide evidence about events that have occurred as the Solar System formed and evolved. Most meteorites (96% of all falls) are stony, and are made up of the same silicate minerals as many terrestrial rocks. The stony meteorites are subdivided into chondrites and achondrites. The former are meteorites that have remained unmelted since formation (or aggregation) of their parents, whereas the latter are igneous rocks, such as basalts, that formed from melts on their parent bodies. As a consequence of the melting process, achondrites no longer have the same composition as the dust which formed the Solar System.

Chondrites retain a chemical signature close to that of the original material from which they aggregated. Although these meteorites have not melted since their formation, they do contain materials that were once molten, and were produced by rapid cooling of droplets of molten stone. The droplets came from collisions between clumps of dust grains in the early stages of the formation of the Solar System, so meteorites such as these represent the materials from which the Solar System grew.

Chondrites also contain organic compounds in varying quantities. It is meteorites like these, together with the ice-rich and organics-rich comets, that probably brought volatile materials to the newly-formed Earth, and helped establish our planet's atmosphere and oceans. Without them, there would be no life on Earth.

The second large division of meteorites, the irons, consist predominantly of iron metal, typically with 5–15% by weight of nickel. These meteorites have all been formed during extensive melting processes on the parent bodies from which the meteorites originated. The heat source for melting was, in some cases, the result of impacts, but for many iron meteorites the heat source most probably came from the decay of short-lived radioactive isotopes, such as ^{26}Al. Iron meteorites are the closest physical analogy we have to the material which forms the Earth's core.

The third and final type of meteorite is that of the stony-irons: a mixture, as the name suggests, of stone and metal. They were also formed by melting in their parent, and represent an intermediate stage between iron meteorites and stony meteorites — a snapshot of material from the core–mantle boundary of the body.

In addition to meteorites from the asteroid belt, there are currently eighteen meteorites from the Moon in the world's collections. Lunar meteorites can be compared directly with samples brought to Earth by the US Apollo and Soviet Luna missions between 1969 and 1976. The surface of the Moon is covered in craters caused by impacting bodies. If the asteroid fragments strike the surface with enough velocity and on a favourable trajectory, then the force of the impacts will be enough to eject material from the surface with a velocity great enough to overcome the Moon's gravity and be launched into space. Subsequently, the material goes into orbit in interplanetary space, and some of it eventually lands on the Earth as a meteorite. In the same way, rocks have come to us from Mars: we have 14 meteorites that have been ejected by impact from the surface of our neighbouring planet. By studying these meteorites, scientists can learn about events that have taken place in the past on Mars, when it had a thicker atmosphere and could support running water, even though the surface of the planet now seems to be dry. The study of organic compounds and salts produced by the action of water in Martian meteorites has demonstrated that these meteorites contain information that might shed light on the possible evolution of life on Mars.'

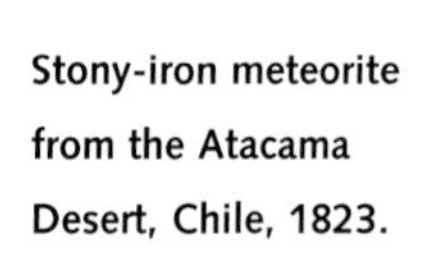

Stony-iron meteorite from the Atacama Desert, Chile, 1823.

EASY ACCESS — ORGANIZING AND COMMUNICATING GLOBAL SYSTEMATIC INFORMATION

Rapid advances in computer technology and the development of global computer networks over the past decade have revolutionized the ways in which we can access and deliver information. Computerized database systems make it easy to search and retrieve information, and using them to create internet sites makes them accessible to millions of people all around the world. At present, in Britain and other countries most of the collections, and the information derived from systematic research, are not computerized; however, the success of a growing number of pilot projects and the need for a global information system to support the international biodiversity initiative suggest that this situation is set to change.

There are enormous benefits to be gained from developing computerized systematic database systems. For instance, although many specimens stored within national collections are located far from their place of origin, computerized databases will provide scientists and conservationists from other countries with immediate access to these, and to related information.

There are two main types of information that researchers can glean from biological collections. The first is detail associated with specimens and samples, and the second is based on the generation of species lists. Although both types of information can be included within the same database they serve different purposes. Specimen data provide information about a particular locality, while species data provide a summary of the taxonomy of groups of organisms. Even in its most basic form, this information is invaluable because it can be used to generate distribution maps and checklists of species. Experts have recently used specimen data, for example, to compile checklists and publish a number of inexpensive plant and animal identification guides that will be used by local scientists to expand understanding of regions rich in diversity, such as Costa Rica and Panama.

This exquisite example of a looper moth is from Costa Rica.

Experts can use this information in other ways too. They can combine species distribution maps, for example, with environmental data from geographical information systems (GIS) to help conservationists and policy makers reach decisions in planning and conservation. The government of Mexico recently gathered data on the distribution of Mexican birds and mammals from specimens and collections in Europe and the USA to help them to plan a national strategy for reforestation, development and conservation.

So why haven't more species databases been computerized? Compiling species databases isn't the simple task that many of us might assume. Malcolm J. Scoble, a research entomologist, explains why.

'Several colleagues and myself recently completed a species database, or Global Taxonomic Facility (GTF) for the moth family Geometridae. This is a large group of insects that includes over 20,000 species worldwide. Their 'looper' caterpillars eat large quantities of green vegetation and the group was therefore chosen because it was both species-rich and ecologically important. Another reason was that, although considerable work still needs to be done on the global classification of geometer moths, the

Museum houses a substantial collection and therefore a rich depository of information about this particular group of insects. It consists of approximately 1 million specimens collected from across the globe, most of them accumulated in the 19th and early 20th centuries.

The main body of the geometer moth collection is arranged taxonomically on a world basis, and is supported by an unrivalled library of taxonomic literature. The arrangement of the collection is mirrored in a card index to the level of species and subspecies, and it was the data on those cards (updated with recent research findings, and information from the original descriptions) that were collated in the database and the resulting published catalogue. Original descriptions are the primary source of taxonomic information, and we knew that it would be invaluable to give users of the catalogue access to the complete chronological range of publications, the earliest of which date back to the 18th century. Finally, we wanted to provide access to a large volume of previously unpublished information that had been added to the manual card index by a succession of curators throughout the collection's long history.

Compiling a computerized database from the information stored within an existing card index is not a straightforward task. One of the problems in taxonomy is that inadvertent duplication of effort can lead to alternative names or synonyms being allocated to the same species. In each case, we had to determine the correct name before we could make a database entry, and this required careful comparative research to delimit the species according to the rigorous application of the internationally agreed rules of nomenclature. Furthermore, many of the index cards have been written by hand and were difficult to decipher without a specialist knowledge of the field.

The GTF took three years to complete. It includes records for approximately 35,000 names, not only the names that have been validated for each species, but also the synonyms that arise when they have been described more than once. We benefited from having a first database, which had been developed by one person over approximately 8 months for another project to study the patterns of species-richness in the Geometridae. Adding and interpreting extra data sets and fields took another 20 months, while the remaining 16 months were devoted to checking the details of the database and resolving outstanding queries.

The aim of the GTF is to give users a system that provides a far more efficient way of accessing and retrieving information than the traditional system of physically thumbing through the extensive card index. It was also intended to show that, in general terms, large quantities of biodiversity data stored in museums can be made more widely available, and in a relatively short period, by a team approach.'

Collection of geometer moths consisting of over 1 million specimens collected from all around the world.

Why Should the Museum Have it all?

Now that we have looked in more detail at some of the ways in which scientists are using the collections to advance our understanding of the natural world, it will be interesting to revisit some of the more controversial aspects of collections and natural history science.

COLLECTIONS OF HUMAN REMAINS

In the previous chapter we saw that experts follow strict international codes of practise and guidelines on collecting specimens abroad, and that after they finish their research, they undertake to return one of each species, including every type specimen, to its country of origin. But what about the specimens that were collected abroad, long before these rules and guidelines existed? Controversies, such as the Greek government's campaign to have the Elgin Marbles returned to Greece from the British Museum have attracted considerable debate over recent years, and none more so than whether or not the The Natural History Museum should continue to house human remains. Paul Henderson, Director of Science, explains why.

'Our own origins and development have probably always intrigued us. The quest for the truth is vigorously pursued by different individuals, groups and professions — not least by anthropologists. The tools that various people use may differ; for the anthropologist, the collections housed in museums are often vital.

The Natural History Museum's research in biological anthropology is based largely on the Museum's existing collections, but also on field studies, when more specimens must be collected and studied, to help meet new problems and challenges. The work centres on the evolution of human beings and considers the broad context in which that evolution has taken place.

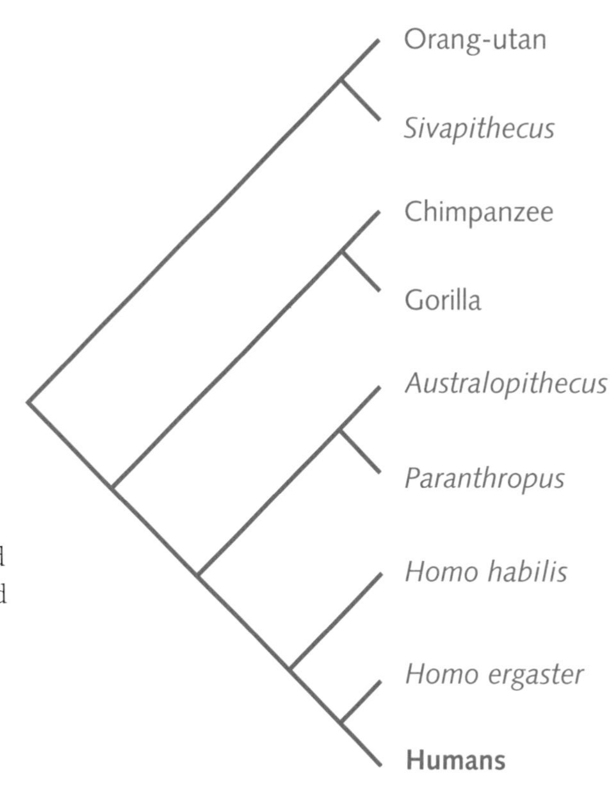

Cladogram illustrating the relationship between humans and primates. The closest living relatives of modern humans (*Homo sapiens*) are the chimpanzee and gorilla.

All creatures great and small are interrelated in some way. The nature of the relationships can be depicted in the so-called 'tree of life' that shows the different ancestors of the numerous species on the various limbs and branches, all of which have arisen by the process of evolution. In this tree, humans and primates are as important as any other group of organisms. Humans are part of natural history and we cannot understand one part properly without understanding the others.

Our Museum collections of human remains are essential to this pursuit. Studies of the shape and properties of these remains enable us to determine the relationships between different groups, between different genera and between different species. In addition, the anatomy of

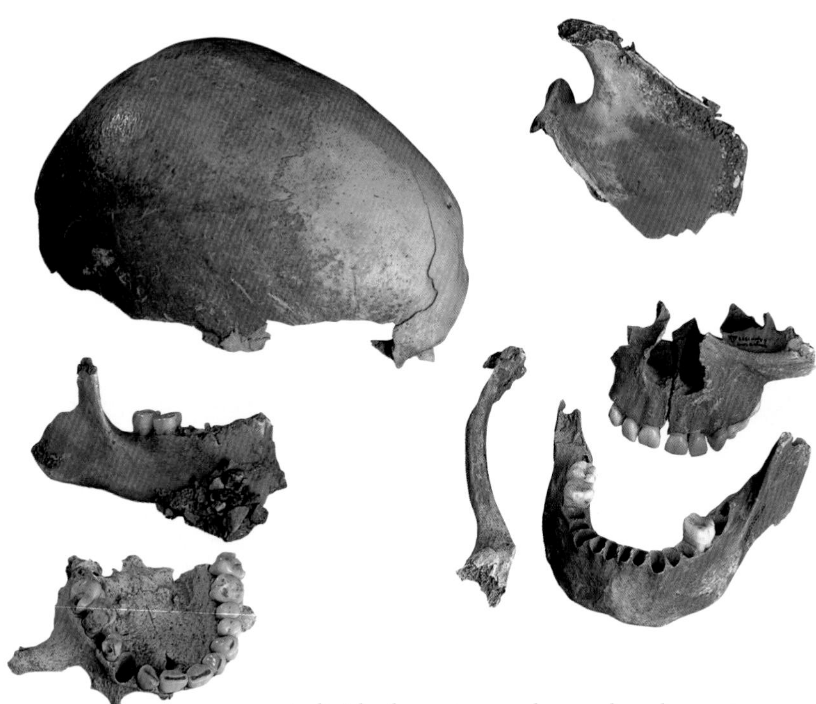

Skull bones, about 9000 years old, from Cheddar Caves, England.

individual species can be used to determine the detailed functioning of the individuals.

Reliable knowledge of the world comes from the vigorous validation of theories and conclusions through experiments and observations. In some areas of biological anthropology, the desired level of knowledge may be delayed for a while because of not having *enough* examples to examine. It is thus important for the collections to be as wide-ranging as possible — especially with respect to age in the fossil record, geographical range and type of habitat.

The geological record, even for more recent times, does not yield up its evidence readily. Details of where mankind and our ancestors were and at what time have to be gained in painstaking ways.

The Museum has been working since 1983 at Pasalar, in Turkey, where half a hillside has been dug away to collect many hundreds of fossil hominids. If one considers that the work entails the careful removal and sieving of sediment or rock then, from the huge scale of some excavations, one can appreciate the great care and time that has been spent.

Male thigh bone stained red with ochre, about 26,500 years old.

Remains of human beings — in rocks, sediments or soils — are far from common because geological processes tend to remove them. Collections in museums often require decades of work to produce a really useful research resource. That building process is, of course, still continuing as more evidence and new ideas unfold. It is also vital to test ideas. For example, the 'out of Africa' hypothesis, which argues that our species, *Homo sapiens*, had a relatively recent African origin, is supported by an increasing number of specimens from different parts of Europe and elsewhere over the geological time of 30,000-100,000 years.

The usefulness of the collections extends beyond learning about human evolution in broad terms. They help provide answers to numerous other questions. These include questions about diet, where the wear on teeth can provide clues, as can the concentrations in the bone material of some indicator chemical elements. Or they may solve puzzles about stature and general state of health, where the size and shape of bones, with or without evidence of distortion caused by disease, can help. Questions about social behaviour may be answered from the grouping of bones of humans and animals in a location or from evidence of practises, such as cannibalism, which can inflict a characteristic type of damage to the skeletal remains. And questions about the longevity of past peoples and how this links to the conditions of the time can be tackled, using evidence from both the fossilized human remains and the nature of the geological deposits associated with them.

There are numerous reasons why we want answers to these sorts of questions — ranging from the pure intellectual pursuit of knowing about ourselves and our environment to using the evidence of the past to help tackle important issues of today, including human health.

All collections in the Museum are professionally curated and protected to a high standard. Collections of human remains are, however, treated with a particular respect. Holding such collections can be a sensitive issue since different cultures can perceive this matter in very different ways. The collections are only accessible to *bona fide* research scientists who are studying human origins and variation.

Some people argue, however, that plaster casts or detailed photographs should be all that is necessary to keep in a museum and that the original material should be reburied. These alternatives have been shown repeatedly to be unsatisfactory for much scientific work, especially as new methods are initiated or new lines of evidence start to emerge. One powerful example of this is the recent work done on the skeleton of a Neanderthal man. From this specimen some DNA was extracted. The nature of that DNA showed clearly that Neanderthals are on a different evolutionary branch from that of our own species, so helping to resolve an important debate about mankind's precursors. No plaster cast could have provided that kind of information, and no one would have thought at the time of collection of the Neanderthal specimen that it could have been used in this way. Research directions and methods are often far from predictable.

The Museum's storerooms may be dark, but the specimens within them, now and at unknown times in the future, will throw light on some of the most fundamental questions of evolution, and on humanity's adaptation to changing environmental conditions. We believe we have a responsibility to provide the right tools for the scientists of the future as well as for those of today.'

Main excavation area at the hominid site, Pasalar, Turkey.

YOU CAN OWN YOUR OWN FOSSIL BUT DO YOU REALLY WANT TO?

If there are strong scientific arguments in favour of retaining important specimens to help experts to increase our understanding of the natural world, there is an equally compelling argument against privately owned fossils. The demand for privately owned fossils has increased dramatically over the past twenty years, and professional collectors and dealers are setting and realizing increasingly high prices. This is becoming a serious problem for public institutions, like The Natural History Museum, that find it increasingly difficult to compete. If the trend continues, fewer specimens will find their way into public museums, and experts are becoming increasingly concerned about the effect this will have on the future development of science. Angela Milner explains why:

A DINOSAUR NAMED SUE

At 10.15 am on 4 October 1997, one of the most complete *Tyrannosaurus rex* specimens ever found fell under the gavel at Sotheby's auction house in New York City. Nicknamed 'Sue', after Sue Hendrickson, the women who discovered her, it was found in a river bank in South Dakota in 1990. Hendrickson was one of a team of fossil hunters working for The Black Hills Institute, commercial fossil dealers who paid Maurice Williams, a Sioux, who owned the land on which the *T.rex* was found, $5,000 for the specimen. The dealers then proceeded to collect the fossil and transport it to The Black Hills Institute, where they began some of the initial preparatory work.

It turned out, however, that because the US Department of Interior held William's land in trust, he had to gain federal permission to sell the specimen, and federal agents seized the partially prepared specimen so that the true ownership of *T.rex* could be established. A long legal battle ensued, The Black Hills Institute lost their claim, despite having spent over $200,000 on collecting and preparing the specimen, and Williams and the government eventually agreed to sell the dinosaur at auction.

The bidding began at $500,000. Eighteen minutes later, it had reached $3.5 million, and, finally, at 11.24 am 'Sue' was sold to The Field Museum, Chicago for $7.6 million. With the buyer's premium, the total price added up to a staggering $8,362,500, most of which went to Maurice Williams after Sotheby's received its commission. No money was paid to The Black Hills Institute.

'Sue' did finally end up in a public museum, but only with the financial backing of a long list of contributors, including the McDonald's Corporation and Walt Disney World Resort. The public will be able to see the fossil being prepared in the newly built McDonald's fossil laboratory at The Field Museum.

Partly prepared skull of 'Sue' prior to full preparation and restoration by The Field Museum, Chicago.

'Science is a cumulative process. We advance our understanding of the world by adding to the vast body of knowledge built up by our predecessors over hundreds of years. Strict conventions govern the procedure by which we are able to make a contribution, and this ensures that once our findings have been checked and validated by independent experts, they continue to be open to scrutiny by scientists, who may wish to re-examine them in the light of new evidence. In palaeontology, this means that scientists should only publish research on a fossil that is freely available for subsequent study by others.

Dinosaur fossils are a good case in point. They are so rare that even the tiniest fragment may hold vital clues that will help us to understand more about the history and lives of these remarkable creatures. In isolation, they may tell us very little and, as the case of *Baryonyx* demonstrates, it is only when they are placed alongside a growing number of similar discoveries that exciting new pictures begin to emerge.'

Tying it all Together

'Why then is not every geological formation and every stratum full of such intermediate links? Geology assuredly does not reveal any such finely graded organic chain, and this, perhaps, is the most obvious and gravest objection to my theory. The explanation lies, I believe, in the extreme imperfection of the geological record.'

On the Origin of Species, 1859
Charles Darwin

Depending on your point of view, one of the advantages, or disadvantages, of a book like this one, which provides an overview of a vast subject area, is that none of the topics, projects or people mentioned are discussed in any great detail. This final chapter, however, takes a change of pace. It follows one research project from beginning to end, to show how the scientific process works in practice and how the various concepts and ideas, introduced in the preceding chapters, tie together to increase our understanding of a new dinosaur discovered in Surrey.

A spectacular discovery — Bill Walker holding the dinosaur claw bone.

The Discovery

On 1 February 1983, Trevor Batchelor, an amateur palaeontologist, paid a visit to The Natural History Museum for advice on a fossil his father in-law Bill Walker had discovered in a clay pit near Ockley, in Surrey, on 7 January. The fact that he wanted to speak to a palaeontologist wasn't unusual — many amateur enthusiasts bring fossils to museums for identification. But from the intriguing description that dinosaur expert Angela Milner

DINOSAUR FOSSILS

WHAT IS A FOSSIL?

Fossils are the naturally preserved remains of living organisms. Most fossils are formed from bones, teeth, shell or wood, but preserved feathers, eggs, skin imprints, faeces (coprolites), and footprints are classified as fossils too. Fossils are formed when minerals such as silica or calcium carbonate become incorporated in or alter the original mineral (calcium phosphate) of the bone, usually buried beneath the ground. As the illustration below shows, fossils often appear a different colour from the surrounding rock matrix: in this case, the preserved bones appear much darker. The colouration of fossils can vary considerably, and depends on the mineral composition of the surrounding rock, as well as on the effects of heat and pressure within the Earth's crust.

WHY HAVE SO FEW DINOSAUR FOSSILS BEEN DISCOVERED?

The chances of any land animal becoming a fossil are extremely rare, requiring a highly improbable sequence of events. The dead organism must be buried quickly so that it is protected from scavengers and other destructive influences, such as temperature fluctuations, water and wind. And, once formed, even if a fossil becomes entombed in layers of sedimentary rock, it can still become damaged or distorted beyond recognition by tectonic activity distorting the Earth's crust. The chance of damage increases with the age of the rock, and this further reduces the probability of dinosaur fossils, which may have formed over two hundred million years ago, surviving intact.

The rarity of dinosaur fossils is further compounded by the fact that the rocks in which they were formed are buried deep beneath the ground. The specimen found in the Ockley clay pit, for example, dates back 125 million years to the Cretaceous Period, and rocks from this period are normally located 20 or more metres beneath the ground in southern England. Mining and quarrying activities occasionally expose these rocks or, alternatively, they are forced upward by tectonic activity and exposed by erosion. The south coast of the Isle of Wight and the cliffs that form the Dorset coastline are particularly rich sources of fossils in Britain because the fossil-rich rock layers in the cliffs are continually exposed by natural erosion on the seacoast.

The Dorset cliffs provide a rich source of fossils.

Snout of *Baryonyx walkeri*.

WHERE HAVE DINOSAURS BEEN FOUND?

During the Early Cretaceous, the world was a very different place. The climate was generally much warmer, and the continents occupied somewhat different locations from the ones they occupy today. What would eventually become Europe, for example, lay very close to the equator, enjoying a subtropical climate, while Alaska, Antarctica and Australia were positioned close to the poles. Dinosaurs' bones or their fossilized footprints have been discovered on every continent and on every large continental island in the world, including Japan, Britain and Madagascar, and in every kind of habitat, from deserts and swamps to forests, open plains and beaches.

received over the phone from the receptionist, she knew as she made her way to the entrance to meet Mr Batchelor that, if the information was correct, this was going to be a spectacular discovery.

AN EXTRAORDINARY FIND

Recognized dinosaur discoveries in England date back over 170 years, but only limited evidence of meat-eating species exists. The same is true for dinosaur discoveries worldwide. To date, for example, only eleven relatively complete specimens of *Tyrannosaurus rex* have been found. Scientists believe that this is because, just as with large mammal predators today, there were fewer carnivorous dinosaurs at the top of the food pyramid, and therefore less chance of finding fossil evidence of their remains. Until 1983, most British discoveries of carnivorous dinosaurs, or theropods as they are collectively known, have been based on the evidence of a few bones or teeth, and there had only been one significant find, the partial skeleton of *Eustreptospondylus oxoniensis*, made in Oxfordshire in 1871.

But there was no mistaking the significance of the huge fossil claw-bone that Bill Walker's son-in-law brought to South Kensington that day. Measuring 31 cm around its curved surface, the claw obviously came from an unusual, unidentified, flesh-eating dinosaur. The discovery was the cause of great excitement among Museum palaeontologists. But nobody suspected, even after visiting the Ockley pit and finding large pelvic and hind leg bones beneath the surface of the clay, that it would turn out to be one of the most important European dinosaur finds of the century. That was a conclusion that became clear only after the specimen was removed and the years of pain-staking, fossil preparation, description and identification began to unfold.

Collecting a Dinosaur

Angela Milner, together with Alan Charig, her co-researcher on this project and formerly the Curator of Fossil Reptiles, immediately arranged to visit the Ockley clay pit with Bill Walker, and were delighted to discover evidence of further fossilized bone. The geology of the clay pit told them that the specimen was located in rocks from the Early Cretaceous Period that were about 125 million years old. This meant that not only was this the most complete theropod ever discovered in Britain, but also it was the only reasonably complete specimen to date that had ever been found in Early Cretaceous rocks of that particular age anywhere in the world.

Physical conditions made it impossible to collect the specimen until May 1983. It took a team of eight palaeontologists, aided by several volunteers, more than two weeks to excavate the remaining two tonnes of fossilized bones, most of which were contained in nodules of very hard, iron-impregnated, siltstone rock which had been deposited around them. Like the fossilized claw bone spotted by Bill Walker, some fragments were found unprotected within the clay and, unfortunately, some of the bones

Smokejack's clay pit near Ockley.

were inadvertently broken or crushed by heavy machinery before the team was able to collect the specimen.

Although the skeleton was disarticulated over a 5 x 2 m area, the bones lay roughly in skeletal order, from the dinosaur's skull at one end to its tail at the other. The bone-containing area was first photographed, then a detailed map was drawn up, to record the precise location of each fragment. This map would eventually help in the process of identifying individual bones and reconstructing the dinosaur.

It is extremely rare to find a complete dinosaur skeleton, which makes reconstruction a particularly challenging task. The complexity of piecing together what is essentially an incomplete three-dimensional jigsaw puzzle without a picture for guidance is further compounded by the fact that a single site may contain the fossilized remains of more than one dinosaur. And while it may be relatively straightforward to separate the bones from individuals that belong to different species, it can be extremely difficult to spot the difference between individuals, even if they belong to a different sex. It's easy to make a mistake, and occasionally bones from two different species have been fitted together in the same model. In one famous example, a skull from the dinosaur *Camarasaurus*, found lying near a headless *Apatosaurus* skeleton in 1877, was incorrectly identified as the missing part, and *Apatosaurus* was reconstructed and displayed using the wrong head. The error wasn't spotted and corrected until 1975.

Once the plan was complete, the specimen was lifted in blocks or sections. Because of the fragile nature of the fossilized material, these

The site survey map records the precise position in which each fragment of the skeleton was found.

Excavating the dinosaur's remains.

Palaeontologists protect one of the blocks of fossilized material with plaster of Paris bandages, ready for transport back to the Museum.

were encased in cocoons of plaster of Paris or expanded polyurethane foam, for transport. The fossils were unloaded directly into the palaeontology laboratory, where the temperature and humidity are carefully regulated to minimize the possibility of deterioration, so that the meticulous preparation work could begin.

A NEW TECHNIQUE

The iron-impregnated siltstone that encased the fossilized remains of the dinosaur was extremely hard, and attempts to dissolve the matrix with thioglycolic acid proved unsuccessful. So the then Head of the Palaeontology Laboratory, Ron Croucher, together with his successor William Lindsay, devised the novel technique of bombarding the fossils with an industrial shot-blaster to remove the bulk of the siltstone and excess clay. Using the shot-blaster, in tandem with generous coatings of latex rubber to protect any exposed area of bone, proved a great success and dramatically reduced the time taken to remove the bulk of the matrix. Once the majority of the rock had been removed, diamond-coated circular saws, tungsten-carbide-tipped chisels and pneumatic dentist's drills were used to remove the remaining layers. For the most delicate work, close to the surface of the bone, the team used binocular microscopes to magnify the removal of the last remaining grains of rock, using finely pointed jeweller's engraving tools.

Preparation takes a very long time: it can take several weeks, for example, for one person to prepare a single dinosaur vertebra (*Diplodocus* had more than 90 vertebrae and its skeleton includes more than 300 individual bones). And because there can be hundreds of different fossil specimens awaiting preparation at any one time, Museum palaeontologists rarely have the time to prepare every part of a single specimen. Instead they prioritize, beginning with the most critical parts of the skeleton, the diagnostic bones, that give palaeontologists important clues about how the animal relates to other species of dinosaurs. Each main group of dinosaurs has a unique skull structure, for example, so even the tiniest fragments of skull can help palaeontologists piece together an understanding of where a specimen might fit in the dinosaur family tree. Teeth provide information about the animal's diet, and the vertebrae reveal important details about its anatomy and locomotion.

BEHIND THE SMILE OF A CROCODILE

Work on the new dinosaur began with the skull. Angela Milner and Alan Charig knew from the exposed claw bone that the animal was a flesh eater, but in addition to telling them more about its feeding habits, they hoped the skull would give them important new information about its owner's relationship to other known theropods. The team was only able to recover parts of the skull, but as they started to work on what appeared to be the largest of these fragments, they realized that Bill Walker's find was even more remarkable than they had at first imagined. Working slowly and meticulously, Ron Croucher and William Lindsay stripped away the surrounding rock matrix to reveal a strange snout and a lower jaw that appeared very different from that of

Spinosaurus was so markedly different from every other known species of theropod in the early 1900s, that Stromer designated it a member of a new lineage of specialized theropods called the spinosaurs, in recognition of its distinctive neural spines.

every other known theropod species — except one, the extraordinary *Spinosaurus aegyptiacus*. This
was discovered in Egypt in 1901 by German palaeontologist Ernst Stromer and its scientific description was published in 1915.

Ernst Stromer's entire collection was destroyed when the Munich museum in which it was kept was bombed during the Second World War, and only his detailed description of *Spinosaurus* remains. The partial fragments were found in rock that was approximately 95 million years old, and included part of a long, slender lower jaw, several conical teeth, and a number of vertebrae with unusual elongated neural spines that measured up to 1.69 m in length. These spines would have created the impression of a sail, standing erect on the dinosaur's back in much the same way as the elongated vertebrae of another, much earlier, mammal-like reptile called *Dimetrodon*.

The new dinosaur's skull measured 90 cm in total length, and the snout was exceptionally long and low — far longer than the compact, deep skull characteristic of other large theropods like *Allosaurus*. One of the larger fragments consisted of the end of the snout, and part of the upper jaw still retained a few of its teeth. The left-hand side of the lower jaw was much more complete, and showed that, despite having snapped in two, it contained an unusually large number of teeth, 32 in total. The two fragments from the lower jaw fitted together perfectly, and when they were placed side by side with the fossil snout, the resemblance to the jaw and snout of a familiar modern reptile was uncanny. 'This is no dinosaur,' several international visitors to the palaeontology laboratory were heard to observe, 'these fragments are from the jaws of a crocodile'.

Having examined its claw bone and other exposed parts of the skeleton, Angela Milner and Alan Charig were certain that they were working on the skull of a dinosaur, but its superficial resemblance to a crocodile encouraged Angela to pay a visit to the

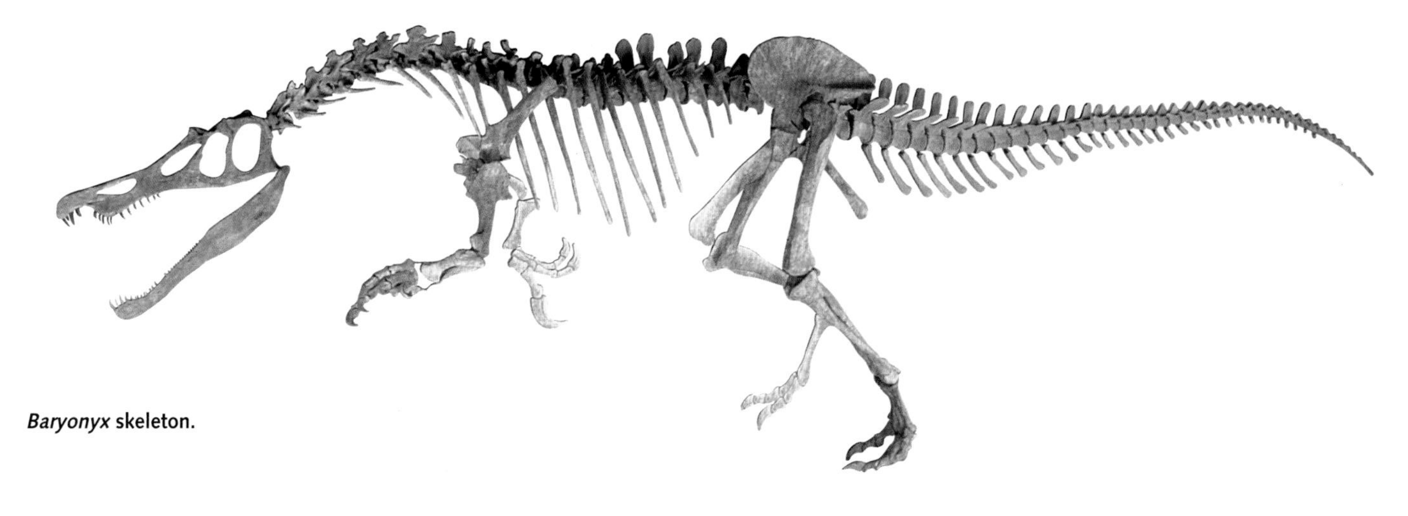

Baryonyx skeleton.

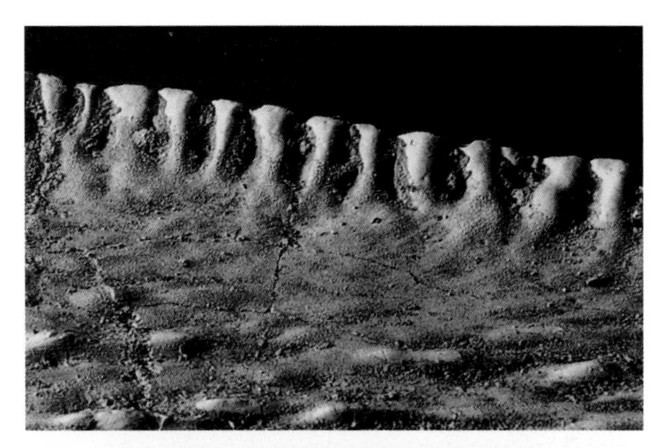

SEM of a tooth from *Baryonyx walkeri* shows fine tooth like cutting edge.

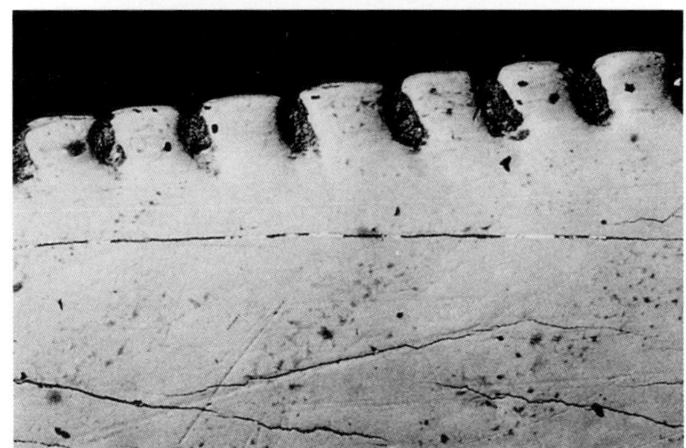

SEM of a tooth of a typical theropod dinosaur, *Megalosaurus dunkeri* with large tooth like projections on the cutting edge like a steak knife.

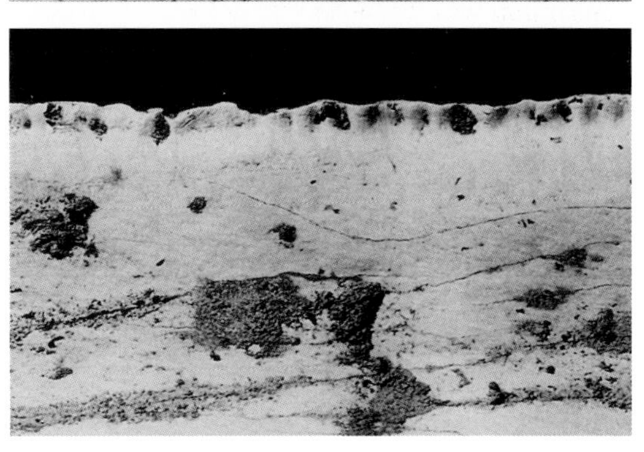

SEM of a tooth of a contemporary crocodile, *Goniopholis* with a smooth, sharp cutting edge.

Museum's modern reptile collection. She wanted to make a series of detailed comparisons between the jaw structure of the new dinosaur and different species of crocodiles, to see if this might help her to understand more about the lifestyle of the ancient theropod.

'When the preparation of the skull fragments was complete and we were able to reconstruct the skull, a series of detailed comparisons with contemporary fish-eating crocodiles confirmed that there was a striking similarity in the s-shaped curve of their jaws. This new dinosaur had twice the number of teeth of any known species of theropod, and its teeth were much more like those of a crocodile than other theropods. These observations led us to hypothesize that its diet may have consisted of fish as well as more conventional 'meat', and detailed comparisons of its teeth with those of typical theropod dinosaurs and contemporary crocodiles seemed to support this argument. Crocodiles have conical teeth that are ideal for skewering fish, completely different from the typical 'steak-knife', cutting-and-slicing teeth found in most theropod dinosaurs and, although not identical, the new dinosaur's teeth appeared more like those found in crocodiles. With its very similar adaptations — pincer-like jaws and stabbing marginal teeth to help grasp, manoeuvre and swallow slippery prey — it is likely that this new dinosaur ate in a very similar manner to modern crocodiles. It probably seized fish sideways across its jaws, before tilting its head backward to move the fish into position so that it would slide headfirst down into its gullet.

As preparation continued, we found more evidence to support our fish-eating hypothesis. Over 10 m long, and weighing approximately 2 tonnes, the new dinosaur would have stood between 3 and 4 m tall on its hind legs. It had extremely powerful forearms and at least one enormous pair of talons on its fingers that were probably used to catch and dismember its prey. Techniques may have included 'gaffing' — hooking or flipping fish out of the water — in much the same way as North American grizzly bears catch salmon today.

Up to this point, however, the evidence for our hypothesis was purely circumstantial, and it was only a year later, when the team began work on the dinosaur's smashed-up ribcage, where its stomach contents were once located, that we found firm evidence to support our

'TOADSTONES' — LEPIDOTES TEETH

'Sweet are the uses of adversity,
Which, like the toad, ugly and venomous,
Wears yet a precious jewel in his head...'
Act 2, Scene 1, *As You Like It*
William Shakespeare

Lepidotes had unusual button-like teeth that studded the inner lining of its mouth, rather like the marbles embedded for decoration in dishes and ashtrays that were popular during the 1970s. This irregular scattering of teeth was ideally adapted for grinding and pulverizing the shellfish that formed the staple part of its diet. According to medieval folklore, *Lepidotes* teeth were magical stones that formed within the heads of toads and were used for medicinal purposes as an antidote against poison and a treatment for epilepsy. This interpretation might seem bizarre today, but without a living equivalent of *Lepidotes* in England during the Middle Ages, there was no reason why anyone picking up a fossil that looked like a rounded stone should connect it with the teeth of a fish. Palaeontologists still face exactly the same difficulty today when attempting to identify bizarre fossilized organisms, or fragments of organisms, that bear little resemblance to known species of plants or animals, or whose appearance may have altered profoundly during fossilization.

The button-like teeth of *Lepidotes*.

ideas. Preparation revealed remains of the dinosaur's last meal — bones from a juvenile plant-eating dinosaur *Iguanodon* and the partially digested scales and teeth of the common Mesozoic fish *Lepidotes*. It may seem surprising that a dinosaur that weighed as much as 2 tonnes could survive mainly on fish, but it is important to remember that many ancient species of fish were much larger than those that populate freshwater lakes today. Some species of lungfishes, for example, grew to lengths of over 5 m during the Cretaceous Period.

A feast for a dinosaur; fossilized teeth and scales identical to this contemporary fish *Lepidotes* were found in the area where the *Baryonyx* stomach would have been.

Everyone wants to know what colour the dinosaurs were, what noises they made and how they cared for their young but, short of discovering living dinosaurs in some unexplored region on Earth, there are many aspects of their life that will remain a mystery. Fossils will only ever tell us part of the story, and the experts can only make reasonable assumptions to fill the gaps. It isn't beyond possibility, for example, that the dinosaur from the Ockley clay pit could swim. There are no special anatomical adaptations like webbed feet to support this hypothesis, but equally there is nothing to contradict it either. Nothing in the skeletal anatomy of humans and grizzly bears suggest that they can swim, but we know they can, so why not a hungry dinosaur looking for fish?'

BUILDING UP A PICTURE

While a picture of the dinosaur began to emerge from its fossilized skeleton, Angela and Alan were piecing together an understanding of the environment in which it lived, by analysing fossils in the collections and referring to published papers on the sedimentology and palaeontology of Wealden rocks like those at Ockley. The stratification of the sedimentary rock at the level where remains of the new dinosaur were found showed that it lived in a mudplain environment, including areas of shallow waters, lagoons and marsh. Fossils found in the same general area gave some idea of the plants and animals that shared its habitat. The vertebrate fauna consisted of sharks, bony fishes, crocodiles, pterosaurs and other dinosaurs, including small sauropods and two species of *Iguanodon*. There was also a wealth of invertebrates, including a huge variety of insects and several species of freshwater shellfish. Common plants included the fern *Weichselia reticulata* and the herbaceous marsh-dweller *Bevhalstia pebja*, and there were also filicopsid ferns, horsetails, club mosses and conifers.

Evolution — the History of Life

You may not have stopped to think about it, but if you look at the skeleton of a dinosaur you will see that we share many features in common. We both have four limbs, fingers and toes, for example, which suggests that back in the distant recesses of geological time, we shared the same common ancestor. The reign of the dinosaurs was an important stage in the history of life on Earth, and one of the key motivations of the palaeontologists who study them is to reconstruct their most likely evolutionary pathways.

The Wealden, Surrey 124 million years ago. The new dinosaur, *Baryonyx walkeri*, searches for fish whilst *Brachiosaurus* browses amongst the trees and *Polacanthus* walks up the river bank.

IMPERFECT RECORD

With only a patchy fossil record at their disposal, palaeontologists can't tell us the precise sequence in which individual dinosaur species have evolved. Darwin wrote at length about why there were imperfections in the animal and plant fossil record and today, over 100 years later, it still remains incomplete, despite many new discoveries.

Series of fossils that provide a visible record of new species branching off from older species are particularly rare and some contemporary palaeontologists believe that this is because, rather than occurring gradually over a long period of time, evolution tends to occur in fits and starts. This modification of Darwin's theory of evolution by natural selection is called punctuated equilibrium, and suggests that the changes that lead to the formation of new species from old species have occurred during relatively short bursts of geological time, that punctuate much longer interludes of stability.

CRETACEOUS SUBURBIA

If a 20th-century time traveller paid a visit to the South of England during the Early Cretaceous Period, they would experience a very different environment from the one we know today.

About 120 million years ago, Britain was still part of the European continent, and large sections of Surrey, Sussex and Kent were covered by the Wealden Lake, a vast freshwater lake and huge flood plain that extended as far as Hampshire. The climate was subtropical, similar to the conditions found today in hot, humid, regions like the Florida Everglades, and, although many of the animals may have been familiar, the vegetation would have looked very different indeed.

Most modern animal groups originated before the Early Cretaceous, but the lack of fossil evidence suggests that even if they were present, many flowering plants (angiosperms) remained a rarity. During the Jurassic and the Cretaceous periods, the world's forests and plains consisted mainly of gymnosperms, plants that produce seeds but no flowers, including coniferous trees, cycads and ginkgoes.

The flowering plants may have evolved over the last 100 million years to become the most dominant land-based plants, but conifers still form vast, dense forests in the northern parts of the world, providing a living reminder of how the world might have looked through the eyes of the dinosaurs.

Gymnosperms such as these *Sequoiadendron*, include the tallest trees on Earth, reaching heights of 80 m. Before it became a protected species in the late 19th century, it took teams of nine men 12 days or more to chop these magnificent trees down by hand.

Another gymnosperm, the bristlecone pine tree, *Pinus aristata*, of the Sierra Nevada, USA, holds the record for the plant with the longest life span. The oldest individuals are thought to be around 4900 years old.

Whatever the reason for these discontinuities in the fossil record, most palaeontologists agree that we will never have a complete physical record confirming the precise descent of every species that has ever lived since life began.

However, by using a technique called cladistics, devised by German entomologist Willi Hennig in 1950, palaeontologists have been able to suggest the order in which they think that different *groups* of species may have emerged. Each group is defined by a set of shared characteristics, or anatomical features taken exclusively from their fossilized teeth and bones, and by careful comparison palaeontologists have attempted to establish the sequence in which new features have appeared.

The result is a tree-like structure, called a cladogram, where each branching point represents the appearance of a new character and all the groups above each point represents the species that possess it. For example, *Tyrannosaurus rex* and its relatives belong to the family Tyrannosauridae, whose members share several characters in the structure of the skull, neck and pelvis. Tyrannosauridae also belongs to the larger, more general group Coelurosauria, and an even wider group, the Tetanurae. Species belonging to the Tetanurae share the loss of the development of fourth and fifth digits, and a unique bony extension that attaches one of the ankle bones to the tibia.

This relatively new method of classification has produced some interesting new results, which have challenged some of our traditional assumptions about the evolution of certain organisms. Analysing the features that modern animals share in common with the dinosaurs, for example, has produced convincing evidence that birds are the living descendants of one particular group of theropod dinosaurs called the maniraptorans and, more specifically, the dromaeosaurs.

WHERE BARYONYX FITS INTO THE DINOSAUR FAMILY TREE

By 1986, three years after the initial discovery, Angela Milner and Alan Charig had examined enough of the new dinosaur to realize that although its teeth and skull were similar to *Spinosaurus*, its lack of specialized vertebrae was enough to justify its classification as a completely new species, genus and family. They announced their discovery in the international science journal *Nature* and officially named the new dinosaur *Baryonyx walkeri* — *Baryonyx* from the Greek, meaning 'heavy claw', and *walkeri* in honour of Bill Walker's spectacular find. But how did this new dinosaur relate to other species of meat-eating dinosaurs, and where did it fit in the theropod evolutionary family tree?

Answering these questions was the task of the next phase of research, which would culminate in the publication of the paper that described the new dinosaur in precise scientific detail. Every discovery of a new species, whether it is a fossil or a living animal or plant, must be accompanied by the publication of a formal paper. This fulfils a rigid set of criteria that defines the genus and the species, so allowing other scientists to understand the find. This includes a formal bone-by-bone description of the complete specimen, a detailed map and a summary of how it was found and prepared, and a series of deductions about the lifestyle, behaviour and evolutionary significance of the dinosaur. The paper must be read and approved by a panel of independent experts before publication, and only then are the full results available to the scientific community. This painstaking process of description, evaluation, and validation can take an extremely long time. The paper that described *Baryonyx walkeri* in full, for example,

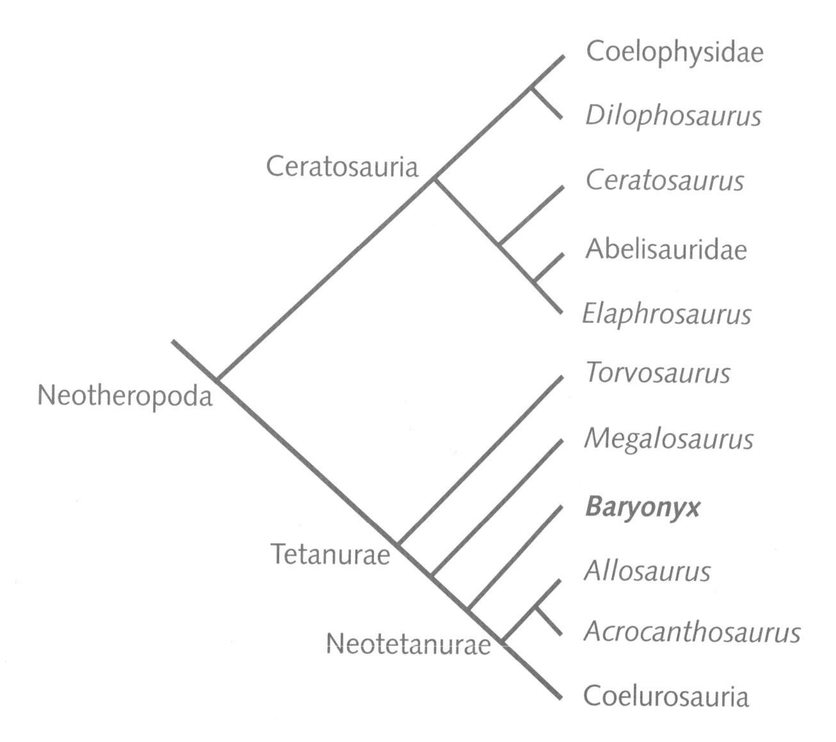

Cladograms are branching diagrams depicting relationships between organisms. This one of the theropods, suggests that *Baryonyx* is most closely related to 'advanced' therapod groups, Neotetanurae and Coelurosauria.

was published on 26 June 1997, a full 16 years after its fossilized remains were first discovered.

Using a range of characteristics to test the relationship between *Baryonyx* and other species of theropods, Angela Milner and Alan Charig produced a cladogram, which shows that the unusual fish-eating dinosaur belonged to the same large group, Tetanurae, as species like *Tyrannosaurus* and *Allosaurus*.

The survey included evidence from the fragmentary fossil remains of two new species of theropods, *Angaturama* and *Irritator*, recovered in Brazil in the early 1990s. Although the evidence for *Irritator* was very limited, consisting of a partial skull that lacked the end of its snout, and only a fragment of snout was found from *Angaturama*, it was obvious that the elongated jaws and teeth from both species shared many characteristics in common with *Spinosaurus*, and the three species were placed in the family Spinosauridae. Angela Milner and Alan Charig noted that *Angaturama* also shared several important characteristics with *Baryonyx*, though they weren't able to demonstrate such a relationship with *Spinosaurus* because the relevant parts of the fossil recovered by Ernst Stromer were missing. This new evidence suggested that the family Baryonchidae, including *Baryonyx*, was more closely related to the family Spinosauridae than any other group of theropods, and should be classified as belonging to the same superfamily Spinosauroidea.

Slowly, a picture had begun to emerge of a specialized lineage of long-snouted, fish-eating predators that lived between 125 to 95 million years ago in Africa, Europe and South America. But how did they evolve and what role did they play in Mesozoic ecosystems? Experts are hoping that two new species of spinosaurs discovered in 1999 will help them to fill in some of the important gaps in our understanding of these extraordinary creatures. The first species, found by American

Reconstruction of *Baryonyx*.

PRIMITIVE BIRDS — THE DINOSAURS THAT GREW FEATHERS

Thomas Henry Huxley first noticed the similarity between the turkey drumstick and the fossilized remains of *Megalosaurus*, reputedly as he was carving Christmas lunch. He found further compelling anatomical evidence to support this in Richard Owen's description of *Archaeopteryx lithographica*, the famous pigeon-sized creature with the feathers and wishbone of a bird and a long reptilian tail, discovered in Bavaria in 1861. However, because no dinosaurs were known to have a wishbone, palaeontologists preferred to believe that rather than originating from dinosaurs, birds evolved from a group of primitive tree-climbing reptiles that learnt to glide.

With no new discoveries to fill the yawning gaps in the fossil record either before or immediately after *Archaeopteryx*, nobody thought to challenge this until the late 1960s. John Ostrom, an American palaeontologist, described a small theropod dinosaur called *Deinonychus*, or 'terrible claw', after the enormous hooked claw on each foot. Ostrom thought the anatomy of the fore and hind limb of *Deinonychus* was very similar to that of *Archaeopteryx*. He visited several European museums where he made detailed comparisons with the specimens to test his ideas. Ostrom concluded that *Archaeopteryx* shared so many features in common with *Deinonychus* that the two had to be close relatives.

Deinonychus.

In 1986, American Jacques Gauthier included Ostrom's findings in an analysis of the evolutionary relationship between theropod dinosaurs and birds. He suggested that birds are most closely related to a group of theropods that includes *Deinonychus* and *Velociraptor*. Since then, there have been a number of important new discoveries which have produced further compelling evidence to support this view. In 1996, for example, the 120-million-year-old remains of the theropod *Sinosauropteryx prima*, found in Sihetun, China suggest that its body was covered with filaments that experts believe could be the early forerunners of feathers.

***Confuciusornis sanctus* male and female bird from Sihetun, north-eastern China.**

Sihetun is particularly rich in fossils because it marks the spot where a lake, once teeming with life, was devastated by a volcanic eruption more than 120 million years ago. Other spectacular discoveries made in this region in the past six years include several more theropod dinosaurs. Some have forelimbs fringed with what might be the forerunners of flight feathers, and *Confuciusornis sanctus*, the earliest known bird to have a toothless beak and a short bony tail. The Sihetun animals show various stages of how feathers may have developed, even though *Archaeopteryx*, which has feathers, is about 25 million years older.

As the anatomical evidence continues to mount, it seems that Huxley was correct and, far from completely dying out, one group of dinosaurs continues to thrive in modern times. They visit our gardens, produce the feathers that insulate our duvets and coats, and provide an important source of food for millions of people throughout the world.

MODELLING

The dinosaur skeletons that you see on display in museums are often replicas, casts of the original specimen made of plaster or bone-coloured glass fibre. Once the modelling process is complete, the next task is to assemble the specimen, in a lifelike pose, which, like most other issues related to the dinosaurs, has been a topic of debate over recent decades. For example, for many years, most experts believed that huge sauropods, such as *Diplodocus* that reached lengths of around 27 metres, were too heavy to support their own weight on land. It was thought they lived an aquatic existence, and the model of *Diplodocus* in the Central Hall of The Natural History Museum depicted a sluggish-looking creature with its vast tail dragging along the floor. When recent research into the mechanics of walking and running were considered alongside fossil footprints that showed that sauropods moved about freely on land, experts realized that this pose was incorrect. Far from dragging on the floor, *Diplodocus* must have held its tail up in the air, as a counterbalance to the long neck and head, so the skeleton was adjusted into a more realistic position. In its new pose, *Diplodocus* looks far more lifelike and the tail assumes a new significance. Besides its role as a cantilever, experts now believe that it may have been used, like a gigantic whip, as a means of self-defence against hungry predators.

Museum experts refit its tail to give *Diplodocus* a more life-like pose.

Reconstructing dinosaurs needn't end with the assembling of their skeletons, and by adding layers of muscles and skin it is possible to gain a reasonably authentic idea of what they would have looked like in real life. The skin texture is modelled on that of modern reptiles and birds and on rare examples of the fossilized imprints of dinosaur skin, which suggests that it consisted of a complex system of nodular scales. Nobody knows what colour the dinosaurs were and, once again, this is an area where points of view have changed with advances in knowledge. For many years, artists coloured their dinosaurs in the dull earthy skin tone displayed by reptiles such as crocodiles, but recently the trend has moved in favour of much brighter colours and more dramatic patterns, in response to the link with birds.

Nobody knows for sure what colour the dinosaurs were but here are a few suggestions from different artists.

palaeontologist Paul Sereno and colleagues in central Niger, West Africa, dates from the Early Cretaceous Period, and represents the most complete spinosaur fossil yet discovered. Named *Suchomimus tenerensis*, which translates as 'the crocodile mimic from Tenere desert', it has an extremely long snout and blade-shaped vertebral spines that, although considerably smaller than those found in *Spinosaurus*, would have formed a shallow sail over the hips.

Combining new information gathered from a comparison of the partial skull fragments of *Baryonyx* and *Suchomimus*, experts have discovered that spinosaur heads were even longer and more slender than previously thought. They have also found that the new species from Niger shares more features in common with *Baryonyx* than it does with *Spinosaurus* or *Irritator*, and have produced a cladogram (right) and phylogram to show the most likely sequence in which the specialized anatomical features of spinosaurs have evolved.

The second most recent spinosaur discovery is currently under investigation by Angela Milner and comes from Morocco. Dating from about 85 to 90 million years ago, the snout from this species is more extreme, longer and more hideous-looking than that of either *Baryonyx* or *Suchomimus*. Based on her observations to date, Angela suspects that it represents a later branch in the evolution of these fish-eating predators. The spinosaurs were only one of several large theropod lineages that lived in North Africa and South America during the Cretaceous. Experts believe that their specialized snouts may explain how they were able to coexist. They lived in coastal or flood-plain environments, and present evidence suggests that they may have evolved in Europe and then dispersed and diversified in the great southern continent of Gondwanaland that existed at the time. Eating fish allowed them to exploit a different part of the food web, in much the same way as crocodiles exist alongside lions today.

Ceratosauria
Neotetanurae
Eustreptospondylus
Torvosauridae
Torvosaurus
Neotheropoda
Tetanurae
Baryonyx
Spinosauroidea
Suchomimus
Spinosauridae
Irritator
Spinosaurus

Cladogram of the spinosaurs. It suggests that the family Spinosauridae includes two closely related pairs of animals, *Baryonyx* and *Suchomimus*, and *Irritator* and *Spinosaurus*.

***Suchomimus tenerensis*.**

The Future

Future Past, Future Present

'Now it is time to expand laterally and to get on with the great Linnean enterprise and finish mapping the biosphere. The most compelling reason for broadening of goals is that unlike the rest of science, the study of biodiversity has a time limit.'

Edward O. Wilson, 1992

A pre-Linnaean view of nature.

In the 18th century, Linnaeus harboured an ambition to catalogue every species of plant and animal on earth and yet today, 250 years and approximately 1.7 million species later, many biologists believe we have still only discovered about one-tenth of the world total. With growing concerns over global warming and the rapid rate of extinction, it's remarkable to think that one of the most urgent tasks facing the world's systematists in the 21st century is the same as the one that Linnaeus set himself in the 18th century.

Charting every species on earth will remain an impossible dream, but charting as much as we can must, and will, continue to remain a critical mission in the new millennium. It's an ambitious plan, not least because of the time it will take to transfer the information that we already have about described species onto electronic databases. Even if it were technically possible, it's anybody's guess, for example, how long it would take to database the 68 million specimens that already exist within the various collections at The Natural History Museum, London and this is only one of the many collections that exist around the world.

So many questions and issues remain to be explored by natural scientists in the new millennium that it would be difficult to provide a detailed summary in the closing pages of this book. Instead, we have approached a number of specialists and asked them to comment on just a few of the areas that are set to puzzle the experts in the 21st century. From the possible threat of doomsday asteroids to the development of organisms and whether or not life exists on other planets, the result is a fascinating glimpse of just a few of the amazing natural science discoveries to come.

Q Will we ever understand how organisms are made?

Per Erik Ahlberg
Researcher, Fishes and Amphibians

'What was my surprise, and I may add, my admiration, in perceiving an ordering that placed under my eyes all the organic systems of this lobster in the order in which they are arranged in mammals?'

Geoffroy Saint-Hilaire, 1822, translation by E.M. De Robertis & Yoshiki Sasai 1996, Nature 380, 37–40

'When studying evolution, we are constantly faced with different body forms that have diverged from common ancestors. On a small scale, such changes are easy to understand — for example, it is easy to see how a mistle thrush could evolve into a blackbird, or vice versa — but the diversity of life also bears witness to unimaginable transformations. We ourselves share common ancestors, not only with the mistle thrush and blackbird, but also, further back in time, with organisms as different as bacteria and trees. How has evolution produced all these different single-celled and multi-celled body forms?

During most of the 20th century, research into this problem focused on natural selection: scientists asked why selection would favour the emergence of particular new structures, such as wings or eyes, and how the earliest and most rudimentary versions of these structures might have functioned. However, this ignores a big part of the problem: how does the organism actually *generate* a new structure? Today, research at the new interface between palaeontology, comparative anatomy and the

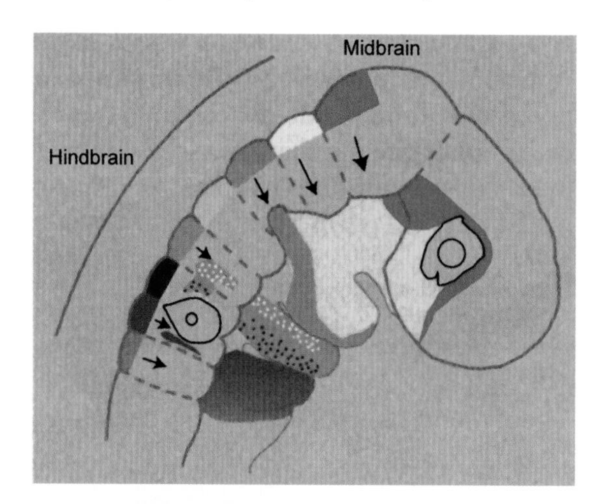

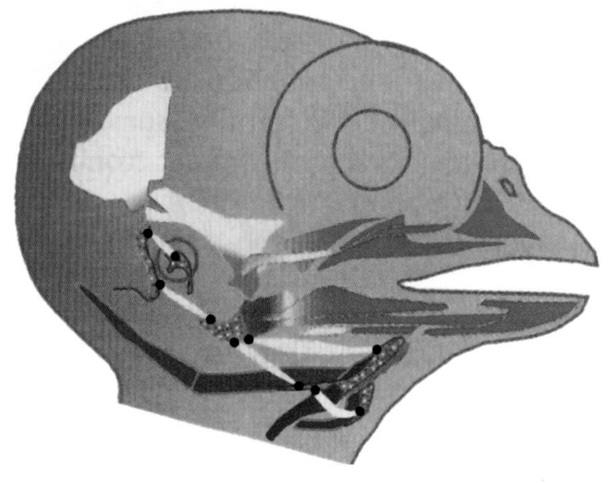

How an animal is made: pictures of a chicken embryo (above) and young bird (left) showing the development of the head. The shape changes dramatically, but the arrangement of segments (shown by different colours) remains unchanged.

ultra-modern field of developmental genetics is beginning to answer this question.

Biologists realized long ago that the instructions for the body plan lie in the genes of the organism; the problem lay in identifying the genes responsible and working out how they steer the development of the body. During the 1980s, geneticists discovered a class of genes, the so-called 'homeobox genes', which are active in the early development of animals and determine where and how different structures develop in the embryo. They found that these genes are similar even in distantly related animals like humans and flies.

The implications of this discovery are astounding. At last, it is possible to compare the body plans of very different animals, establishing which parts are equivalent to which, and inferring how they may have diverged from a common ancestor. Scientists in the 19th century were already tackling this problem, but lacked the tools to answer the questions they raised. In 1822, the French anatomist Geoffroy Saint-Hilaire observed that the body plan of a vertebrate (backboned animal) resembled an upside-down version of an arthropod, such as a lobster or an insect. Most scientists dismissed this idea. Not until 1996 was it shown that the gene activity pattern of vertebrate development is, in fact, upside-down compared to that of arthropods; Geoffroy Saint-Hilaire was right.

Work is underway to tie such genetic discoveries into the evolution of body form revealed by the fossil record. Fossils don't have genes, but they show crucial early stages of the evolution of body form. For example, fossils show that the earliest vertebrate limbs did not have five fingers, but eight; any interpretation of the genetics of limb evolution must take account of this. The coming century will reveal the genetic basis for the evolution of multi-celled organisms, and the emergence of the plethora of body forms that exists in the world today. The dream of the 19th century anatomists will finally have come true: we will understand how organisms are made.'

Q Will scientists discover life on other planets?

Monica Grady

Martian meteorite specialist

'Our planet teems with life, but is it alone in the cosmos? This question has fascinated scientists and philosophers since the astronomer Galileo thought he could see oceans on the surface of the Moon. We know now that we have to look further afield than our harsh atmosphere-lacking partner in seeking life other than in our own oasis of biology. Fortunately, recent space missions suggest there are several possibly habitable niches in the Solar System.

With a supply of liquid water and the presence of carbon, nitrogen and oxygen, even in small quantities, it is possible that organisms could evolve, given sufficient energy and a stable environment. The most likely places to

The continuing quest for evidence of life on the surface of Mars.

look for life are on the bodies that have all these essential ingredients — Mars, the Galilean satellites of Jupiter and, to some extent, Saturn's giant satellite Titan.

Water is present in the polar ice caps on Mars. However, the surface temperatures on the red planet only exceed 0°C during summer at the equator, and the atmospheric pressure is too low for water to be stable at the Martian surface. Evidence that liquid water flowed across the surface in the past is cut deep into the Martian surface in the form of valleys and channels, presumed to have been scoured out by floods. The water was generated from permafrost (the thick ice layers under the surface of Mars) melted by underlying magma. Although Mars probably no longer has active volcanoes, sufficient residual heat might allow water to occur at depth in the crust, providing an energy source for simple organisms. Analyses of sandstones from the Dry Valleys region of Antarctica show that bacteria and algae can survive in conditions of harsh, dry cold, and provide a valuable analogy for potential microorganisms on Mars.

In 1996, features within the ALH 84001 Martian meteorite were interpreted as nanofossils (minute fossils). This report followed the discovery of organic carbon within the meteorite, and led to great excitement worldwide, and a resurgence of interest in the possibility of life on Mars. Tiny fossil bacteria have now been identified in a second of the 14 Martian meteorites, and similar organisms have been discovered living at depth in the earth's crust. Unfortunately, there is no consensus on the interpretation of the features in ALH 84001: many scientists believe that the 'nanofossils' are artefacts. It has also been shown that fossilization processes proceed much more swiftly than had been thought, even at the low temperatures of the Antarctic. Thus it is possible that the meteorite might have become contaminated by terrestrial materials.

Although the Martian meteorites have provided us with much information on the environmental conditions on the surface of Mars, they have not provided conclusive evidence that life has either existed on the planet in the past, or is still present today. In 2003, the European Space Agency (ESA) will launch the Mars Express mission, comprising an orbiter and the *Beagle 2* lander. *Beagle 2* will have on board a fully-integrated package designed to search for the chemical traces of past (and present?) life on Mars, by examining surface and sub-surface soil and rock samples and the atmosphere.

Other possibilities exist further from our planet. Europa is the smallest of the four largest (Galilean) satellites of Jupiter, with a radius a little less than that of our own Moon. It orbits Jupiter at a distance of about 600,000 km, a distance sufficiently close for the satellite to be heated by the tidal forces of Jupiter's gravitational attraction. Most of the information and images that we have of Europa have been acquired by the recent Galileo mission, and so we now have a fairly accurate picture of Europa's density, surface composition, magnetic properties and appearance.

Researchers take a lunch break during their search for meteorites in the Lewis Cliff area of Antarctica.

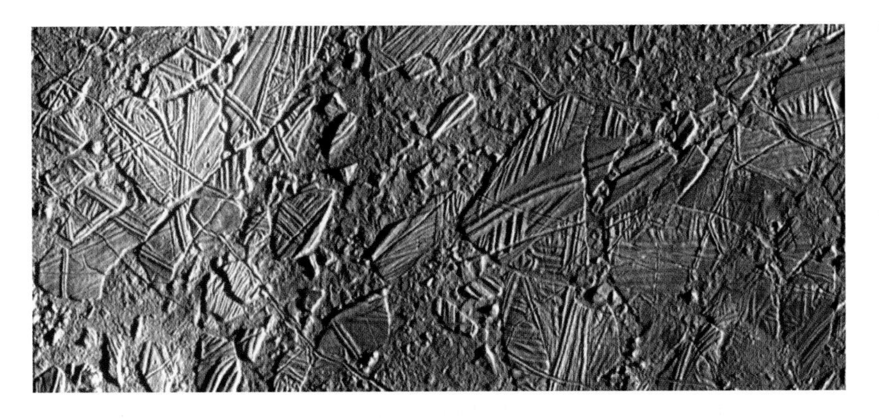

Ice raft on Europa.

Europa's density (about 2.97 g cm^{-3}) implies that it is a silicate body, but with a significant content of water in the form of ice. Recent models based on gravity and magnetic data have suggested a differentiated structure for Europa, comprising a metallic core, overlain by a silicate mantle, over which resides a crust of ice. The icy crust appears to be about 150 km thick, and might also have a layered structure, with a substantial sub-surface salt-rich ocean beneath a thinner shell of ice.

If there is indeed a sub-surface liquid ocean on Europa, then the heat source that keeps it liquid is likely to come from the complex interplay of the orbiting Galilean satellites with Jupiter. Both Io and Ganymede exert minor tidal influences on Europa, in addition to the much larger tidal effect emanating from the parent planet itself. There has been much speculation that Europa's ocean might be heated from the bottom upwards, by hydrothermal vents analogous to those found on the deep ocean floor of the Earth. Study of the sub-surface inland Lake Vostok beneath the ice of the Antarctic plateau is being employed as a pathfinder case study before the forthcoming exploration of Europa and its ocean.

Meteorites are not delivered to Earth from Europa, so the search for life on this icy satellite will have to be achieved by space probes. A first step will be taken by the proposed NASA Europa Orbiter (whose launch is planned for 2003), which will measure the depth of the ice and any water ocean with a radar echo-sounder.

Titan, the fifteenth Moon of Saturn, is by far and away Saturn's largest satellite. The Moon has an atmosphere 1.5 times as thick as the earth's, composed mainly of nitrogen and methane but also containing minor amounts of more complex organic molecules.

Although there is no suggestion that the environment of Titan is in any way suitable to harbour life, the atmospheric composition and surface conditions are primitive enough to have been regarded as possible analogues to the conditions that existed on the early Earth, before the development of the biosphere. So, understanding the atmospheric and surface processes on Titan is a key to understanding the processes that led to the evolution of life on earth. As a consequence of this significance, Titan is the target of Huygens, an ESA-led probe (part of the joint NASA-ESA Cassini-Huygens mission to the Saturnian system, scheduled to arrive in December 2004), which will descend for several hours through Titan's atmosphere. The probe will acquire images, measure the elemental and isotopic composition of the atmosphere, and record its environmental conditions right down to its crash-landing — or splash-landing — on the surface.'

Q Will we be able to address the questions of how life began on earth, whether it developed only once here or whether we will be able to find evidence elsewhere in our Solar System for present or past biological activity on other planets?

Richard Herrington

Researcher, Ore Mineralogy and Economic Geology

'There are conflicting models for how life may have begun on Earth, ranging from seafloor

Fossil organisms preserved around an ancient hydrothermal vent.

hydrothermal vent sites ('black smokers' and other mineral springs) to sites deep within the rocks below the Earth's surface ('the deep biosphere'). Much of the discussion is about how and in what environment the initial organic 'building block' molecules (found in much of the studied meteoritic and cometary material, and therefore likely to have been available on all planets) developed into self-replicating complex RNA and DNA structures.

New analytical techniques mean that research is rapidly converging towards a 'best guess' answer to where life began — or at least to where the earliest recordable life might have developed on our planet. However, this answer will be based only on our experience of studying material here on Earth. Some researchers propose that life may have developed on other planets or satellites within our Solar System, too, since there are places where conditions might have been favourable (such as Mars, Europa and Titan), which may be linked to the development of life on Earth. Exploration of other planets will be able to seek evidence for current or past life there, using techniques developed here on Earth. Studying the likely sites for life on other planets (such as hydrothermal vents and deep water-bearing reservoirs) will contribute significantly to the current debate about the origin of life on Earth, and will yield clues as to the likelihood of life existing elsewhere in the Universe.'

Q What will minerals tell us in the future?

Mark Welch

Researcher, Crystal Chemistry

'There have been significant advances in our understanding of the earth in the second half of the 20th century. Earth scientists have developed the unifying theory of plate tectonics, and advances in computer technology have enabled us to reconstruct the geological history of our planet over the last 2000 million years, and to reveal the detail of the physical behaviour of the mantle and how this drives plate tectonics. Understanding the role of water-bearing minerals in the Earth's tectonic cycle will continue to be a major research topic over the next 20 years, and in the future we will be looking to the Solar System to provide vital clues about the evolution of the Earth. This will include studying tectonic behaviour on planets such as Mars and Venus and on satellites like Titan and Europa, and the role of water, ammonia and methane ices in the evolution of planetary interiors, surfaces and atmospheres.

Two other growing research areas in the earth sciences that will become increasingly important in the 21st century relate to environmental issues. The first involves geophysicists improving their modelling of earthquake behaviour, learning how to improve the accuracy of their predictions, and finding new ways of reducing the effects of major quakes on urban populations. The second concerns the study of reactions at mineral surfaces by mineralogists and geochemists, with results that may enhance the containment and disposal of toxic waste, including heavy metals (such as those from mine waste deposits), radionuclides (from nuclear reactors) and human waste (in landfill sites).'

Q Is it possible to predict where the major scientific advances will be made in the next 100 years?

Malcolm Scoble
Research Entomologist

'It is unwise to anticipate progress in science because individual findings can so radically change the direction and agendas of research. Who could have foreseen, for example, that molecular biology would have been transformed from chemistry to bioinformatics by the discovery of a method to multiply small fragments of DNA for analysis? But it is reasonable to predict that one area to which much further scientific effort will be applied will be developmental biology. Despite impressive findings, we are far from having a true understanding of how the genetic makeup of an organism is translated into its body form.

Yet unless we are more successful in conserving habitats and species, intellectual efforts on many scientific issues will appear increasingly detached from the environment. To succeed even in ameliorating the decline of biodiversity, natural scientists will need to translate far more effectively their findings into the complex mixture of big business, politics, economics, land planning, employment and environmental regulation. Enthusiasm for interdisciplinary work between areas of science has increased. But less has been achieved in injecting natural science into the humanities and environmental decision-making.

Even the most basic of information remains scattered. For example, almost 250 years after the Linnaean dawn of modern taxonomy, we still lack a centralized list of all known species. The development and widespread availability of desktop computers, with ever-increasing storage capacity, has already enabled many scientists to compile large datasets about particular groups of living organisms. Two predictable advances relevant to this area are likely in the new millennium — one is technical, the other organizational.

Firstly, the storage capacity and speed of computers will assuredly continue to grow. Secondly, although less certainly, there will be better organization in the global collation of data on species. Of paramount importance to saving biological diversity will be the human capacity to use the information effectively; but technological advance has nearly always outstripped wisdom. While it is hard to be optimistic about the future of the natural world, natural scientists will be engaged increasingly in the debate about its protection.'

You can see a variety of stony corals together with a number of different species of fish, including basslets, damselfishes, and snappers, in this area of Egyptian Red Sea coral reef.

Helosis cayennensis, **a parasitic flowering plant from the tropics.**

Identifying plants in a Paraguayan research station.

What is floristics and why is it so important for the future?

Sandy Knapp
Research Botanist

'In its simplest terms, a flora is a publication that provides basic information about the plants of an area, including their names, some of their taxonomic history, descriptions of what they look like, where they occur, and a mechanism (a system of keys) by which they may be

identified. All this basic information is typically documented by reference to 'voucher collections' of herbarium specimens and to published taxonomic literature. But why should we continue to do such basic documentary work in these days of molecular biology? What relevance does floristics, the writing of floras, have to tomorrow's world?

As we move into the 21st century increasingly aware of biodiversity and its importance, the role of taxonomy, and by extension, floristics, becomes ever more relevant. Floristic projects have an increasingly important role to play in conservation and systematics, today and in the future. They are the backbone for the generation of collections documenting the geographic distribution of plants and their occurrence in particular areas through time. These collections are also the only way botanists have of documenting the range of variation that exists between and within species. This enables botanists to draw up hypotheses suggesting where the boundaries between different species, genera and other biological groups (taxa) should be drawn, and what their relationships are with one another. Hypotheses about species and their relationships will become more and more robust as more collections are examined, from over a wider geographic range.

Training will always be essential to the future not only of floristics, but also of all of biology. Floristic projects are good training grounds for future botanists, who may not end up working only in the field of floristics, but instead become monographers and specialists researching understudied groups of plants, or work in conservation-related projects. It is easy to train students in one's own speciality and to have them work on an aspect of the family or genus with which one is familiar. The real challenge is to enable a student to become the

Keith Scholey

HEAD OF THE BBC NATURAL HISTORY UNIT

I remember clearly the sense of awe at the age of eight when I confronted my first dinosaur in The Natural History Museum. I was used to encounters with giant creatures as I lived then in Kenya, and had met my fair share of elephants at very close quarters. However, this creature's skeletal frame was so large that it seemed impossible that an animal of this size could ever have walked on the Earth.

A decade or so later this puzzle of mine had grown. As I strove to determine the physical limits of size for flying creatures as part of my PhD thesis, those prehistoric reptiles continued to confound me. This time the giant pterosaurs, weighing in some cases more than twice that of today's largest flying creatures, posed my problem. If the muscles on the bones of those fossil giants were made of the same stuff as today's reptiles, birds or mammals then they should not have been able to take to the air. It seemed highly likely that they were made of the same stuff and so the only way I could get the sums to work to get these giant pterosaurs airborne was to increase the Earth's air density or, more effectively, reduce its gravity! Thoughts such as these were no way to get a pukka qualification like a PhD, so it was better to leave the problem unsolved. Today, whenever I visit the Museum the dinosaur skeleton still haunts me. Twenty or so years on, no one has explained how the pterosaurs might take off and this beast still looks unlikely to have walked on leg bones with the same proportions of an elephant's but miraculously supporting three times the weight. Oh, if only one could change gravity! But mysteries like these are the wonderful thing about The Natural History Museum. Its specimens pose many seemingly impossible questions, some of which have now been answered while still others remain. Its record of bio-diversity drives us to question all aspects of life on Earth — and without it, and without the thoughts of the people who have studied it, 20th-century man would be far less the wiser.

The pterosaur, *Pteranodon*, with a wing span of up to 9 m.

This collection of biodiversity, as it continues to grow, will undoubtedly lead us to more answers about life's history and our own origins. But this understanding also continues to help us to predict life's future. Studies of collections such as these have provided clear evidence that the 21st century will be one of the toughest for life on Earth, but I am certain that the knowledge gained by The Natural History Museum will help to cushion life's roughest of rides through the next hundred years. For this reason alone, the Museum must remain as one of our most precious of institutions. But may it also remain a place of mysteries.

specialist in a group that has no specialist. Floristic projects identify these holes in current taxonomic expertise very effectively, and will enable us to deploy more adequately our scarce human resources to better effect for the future.

Our greatest challenge for the future, however, is to make floristic research timely and accessible to a wide variety of user-groups. It is important that the taxonomic community considers not only purely scientific goals, but also takes into account the needs of nations signatory to the Convention on Biological Diversity. Regional scale projects, such as Flora Mesoamericana, can contribute to national conservation goals, but only if the information is made accessible to a wide variety of users in the widest possible way. Technology, which changes and improves every day, is our most powerful tool for achieving this goal. It is essential that large, long-term floristic projects actively evolve over their life-span. Technology is the greatest ally to improved access and applicability, and we as taxonomists must take advantage of this to the best of our abilities, or run the risk of obsolescence and ultimate extinction.'

Q How can we measure the impact that the continuing loss of biodiversity will have on global ecosystem functioning?

Geoff Boxshall FRS

Merit Researcher, Crustacea

'The wealth of living organisms (biodiversity) provides the global ecosystem with many services that we take for granted, such as the provision of oxygen, the processing of sewage and the sequestration (locking up in a relatively inert form) of excess carbon. It is a surprise to most people when they learn that the oceans provide the majority (about two-thirds) of these global services, but it's true. It is the microorganisms and the lowly invertebrate animals of marine planktonic (drifting) and benthic (seabed) communities that provide us with more oxygen than the Amazon rainforest, remove more carbon dioxide from the atmosphere than all land plants combined, and recycle more raw sewage than all of humanity's processing plants.

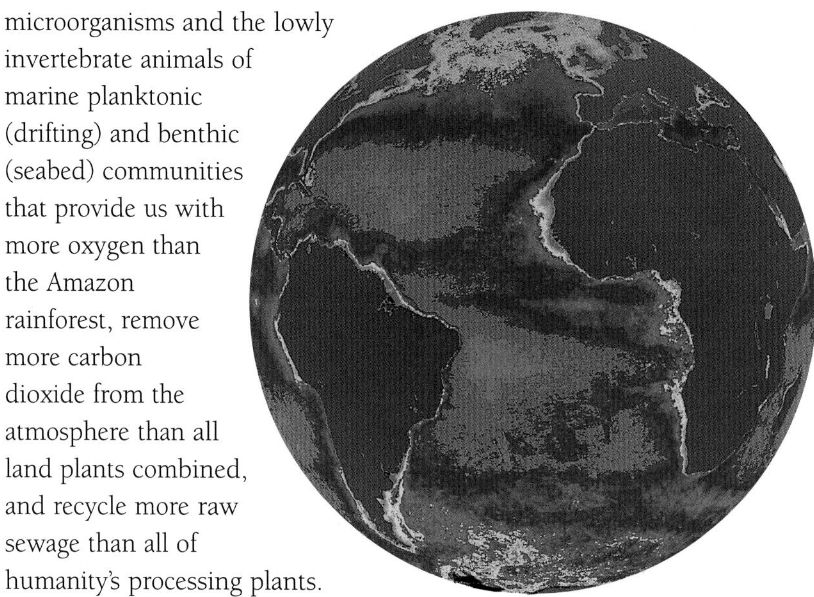

This false colour satellite image of the Atlantic Ocean illustrates the density and distribution of phytoplankton in the surface water, from the densest areas (red) through yellow, green and blue to the least dense areas (pink). The highest densities occur in nutrient-rich coastal waters and as seasonal blooms in the northern Atlantic.

This leads me to my question — if we continue to reduce marine biodiversity by over- exploitation, by pollution and by habitat destruction, will these ecosystems continue to provide such global services? Or, to put it another way, as more species go extinct, and originally complex ecological communities become simpler, at what point do they fail, or at least cease to work in the same way? Answering this question will require a considerable improvement in our knowledge of energy flow through oceanic systems, in our knowledge of marine biodiversity and in our ability to create predictive mathematical models of such systems.'

Q Why must the scientific community engage the politicians in their mission to document the biodiversity of the species-rich tropics?

Richard A. Fortey FRS

Merit Researcher, Arthropods and Graptolites

The problem is that the systematic institutions — and the reference collections — are based in the wealthy western countries, which, historically, have done the job of describing new species.

Equally, the economic potential of living things has been explored by major Western drug companies and the like.

Yet the need 'on the ground' is in poorer countries, such as Madagascar, which have virtually no resources for systematic documentation, and which are also carrying out habitat destruction on an accelerating scale. The human problems of such countries attract media attention and the attention of international political bodies — and rightly so. Yet these problems must be seen as part of an ecological crisis which embraces animals and plants, too. The way forward should be that these developing countries must have a stake in preserving their own biodiversity. The systematic scientist should become as essential to preserving planetary 'health' as any epidemiologist — indeed, with the search for new bioactive chemicals, their paths may cross more and more.

Because species extinction is a global, and moral, problem, it requires international political will, and the collaboration of all taxonomists, to address it. More taxonomists will have to be trained in the places that need them. We don't want this millennium to be one long funeral, interring the richness of the world in a mass, unmarked grave.'

Q Is the Earth safe from the doomsday asteroids and comets?

Matthew Genge

Researcher, Micrometeorites, Meteorites and Petrology

'In the 1960s, we believed in an island Earth whose history was dependent on the planet's internal workings via the newly-proposed plate tectonic theory and was not subject to external influence and disturbance. This reassuring picture was to be shattered by a few scientists who came to realize that certain odd explosion craters found on the Earth's surface were in fact caused by the impact of asteroids and comets. With the revelation that many of these kilometre-sized craters, such as the 50,000-year-old meteor crater in Arizona, were very young came the realization that impacts could still occur at the present time and may pose a significant hazard to life on Earth.

In the 1980s, scientists suggested that the extinction of the dinosaurs and many other species at the end of the Cretaceous Period, 65 million years ago, may have been caused by the impact of a large comet or asteroid with the Earth. This view, although somewhat controversial, led some scientists to suggest that large infrequent impacts were the main cause of global mass extinctions and were important in the biological evolution of the Earth.

Prior to the launch of *Deep Space 1*, which will help NASA track asteroids and comets.

The interest in impacts made astronomers turn their telescopes away from the stars to start looking for asteroids and comets that could collide with the Earth. The first surprise was in 1989, when an asteroid roughly 500 m in diameter passed within the orbit of the Moon but was only observed as it retreated from the Earth. This event, and the discovery that large numbers of asteroids lay in orbits that took them from the asteroid belt and across the Earth's path, compelled many scientists to investigate the effects and frequencies of impacts.

Today, astronomers scan the skies for new near-Earth asteroids and comets, carefully calculating their orbits to determine if they pose a hazard to the Earth. So far, they have discovered only 300 or so asteroids larger than a kilometre in size out of an estimated 1500 that approach the Earth's orbit. Although none of these will collide with the Earth in the foreseeable future, many thousands of large dark extinct comets remain to be discovered, and few of the hundreds of thousands of objects smaller than 500 m have been identified and their hazard assessed.

A comet fragment only 60 m in size was probably responsible for the 12-megatonne explosion at Tunguska, Siberia, in 1908 that devastated 2000 km^2 of forest. This is classed as a smaller event, and similar collisions occur as frequently as every 1000 years or so. Modelling the effects of the impact of small bodies involves knowing the materials that make up asteroids and comets. This has been achieved through the study of meteorites and micrometeorites, which are the small fragments of these objects that are recovered from the Earth's surface. Weak objects such as comets may not collide with the ground to form craters but can cause damage due to their explosive break-up in the atmosphere. Strong objects such as metallic asteroids do reach the Earth's surface, forming craters on land and producing devastating tsunamis on the oceans.

In this millennium, the efforts of astronomers in searching for and characterizing potentially hazardous asteroids will continue, in an effort to provide as much warning of an impact as possible. Scientists at the Museum have been among those working on the effects of the impact of smaller asteroids and comets with the Earth. Such research will allow us to better evaluate both the hazard posed by objects found to be on a collision course with the Earth and the outcome of planet-wide defence strategies that might be employed to mitigate the threat. Although the Earth is not yet safe from doomsday asteroids or comets, scientists are working to protect our world from these natural disasters.'

Simulated asteroid impact with Earth.

SEEING THE LIGHT

The Sun, like all stars, will eventually die. It will exhaust its supplies of hydrogen and helium, expand to up to 200 times its present size, and consume most of the rest of the Solar System, including our planet, in flames. In time, it will cool and shrink, forming a dense

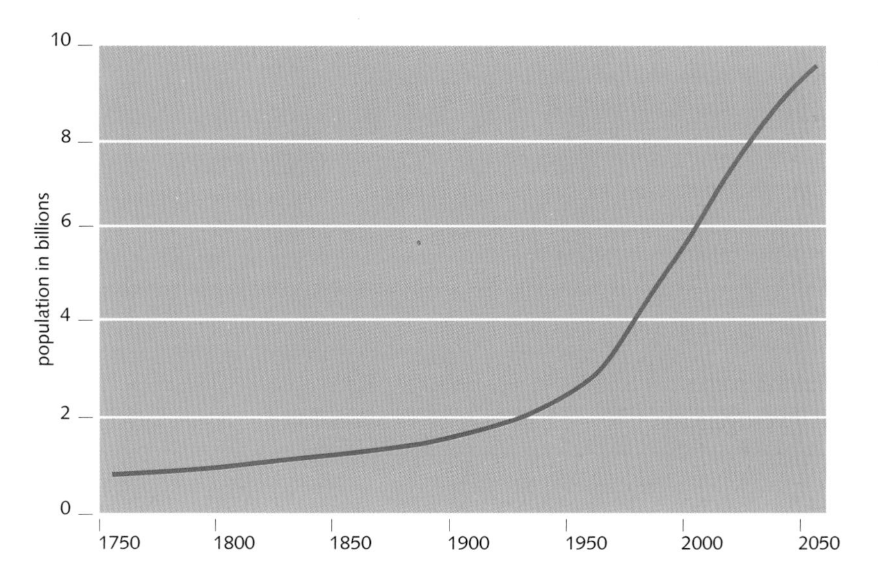

It is estimated that the human population will continue to expand rapidly over the next 50 years.

white dwarf that is roughly the same size as Earth, before dispersing into billions of fragments that will become the building blocks for new stars and galaxies.

Thankfully, most experts predict that this scenario isn't due to take place for another 5000 million years, but the fact that they agree that the Earth will eventually come to an end raises some interesting points. As the new millennium begins, the mystery of the origins of life continues to elude and fascinate us, so it seems ironic somehow that the actions of the world's ever-expanding human populations may be contributing to the annihilation of the very thing that we long to understand. If our activities remain unchecked, if we continue to damage and pollute the environment, accelerate the loss of precious species and influence climate changes, will humans, rather than natural catastrophes like an asteroid impact or the death of the Sun, be responsible for the destruction of life on Earth?

There are many gloomy scenarios, but there is hope too. Hope that comes from the knowledge that scientists have accumulated over the past 300 years spent investigating our planet. Perhaps we will learn from our mistakes, and finally take global responsibility, in deed as well as word, for the guardianship of life on Earth. Perhaps we will discover how to nurture and sustain rather than damage and pollute the environment, so that future generations, like our ancestors before us, may observe the Earth as it continues to change, bathed by the light of the Sun, for many millions of years to come.

Further information

Acknowledgements

With special thanks to Angela Milner, Bob Press and Phil Rainbow for their thorough review of the manuscript, and to Neil Chalmers, the Director, for the support given to this project.
Additional thanks to the following staff of The Natural History Museum for their help with ideas, contributions and images: Per Ahlberg, Peter Andrews, Jeremy Austin, Jane Bevan, Steve Blackmore, Bob Bloomfield, Geoff Boxshall, Martin Brendell, Steve Brooks, Shirley Brennan, David Carter, Barry Clarke, Robin Cocks, Eileen Cox, Gordon Cressey, Oliver Crimmen, Alan Criddle, Steve Culver, Martin Embley, Paul Eggleton, Paul Ensom, Peter Forey, Richard Fortey, Brian Gardiner, Nancy Garwood, Matthew Genge, Mary Gibby, Monica Grady, Peter Hammond, Julie Harvey, Paul Henderson, Richard Herrington, Jerry Hooker, John Jackson, Charlie Jarvis, Chris Jones, Sandra Knapp, Robert Kruszynski, John Lambshead, William Lindsay, Chris Little, Norman Macleod, Colin McCarthy, Christopher Mills, Linda Pitkin, Rory Post, William Purvis, Dave Roberts, Brian Rosen, Malcolm Scoble, Peter Stafford, Chris Stanley, Chris Stringer, John Taylor, Paul Taylor, John Thackray, Eva Valsami-Jones, Roy Vickery, Mark Welch, Peter York and Jeremy Young.

p.159, Jana Bennett, would like to thank Robert Prys-Jones, avian expert, and Julian Hume, research student, Tring.
p.196, picture based on research by Andrew Lumsden and Georgy Koentges, Guy's Hospital.

Further Reading

CURIOUS MINDS

Herbals: their Origin and Evolution, 2nd edn., Agnes Arber. Cambridge University Press, 1938.
Ideas that Shaped our World: Understanding the Great Concepts of Then and Now, Robert Stewart (Consultant Ed.). Marshall Publishing Ltd., 1997.
Science, A. Hellemans and B. Bunch. Simon and Schuster, 1988.
The Compleat naturalist: a life of Linnaeus, Blunt. Collins, 1971.
The History of Scientific Ideas, Steele. Hutchinson Education, 1970.
The Timetables of Science: A Chronology of The Most Important People and Events in the History of the World, Alexander Hellemans and Bryan Bunch. Simon and Schuster, 1988.

TEMPLE TO NATURE

Alfred Waterhouse and The Natural History Museum, Mark Girouard. The Natural History Museum, London, 1981.
Darwin, Adrian Desmond and James Moore. Penguin, 1992.
Evolution, 2nd edn., Colin Patterson. The Natural History Museum, London, 1999.
The Origin of Species, Charles Darwin. Oxford World Classics, Oxford University Press, 1998.

WHAT IS SCIENCE

Earth, Press and Siever. W.H. Freeman and Company, 1997.
The Double Helix, James D.Watson. Penguin, 1970.
This is Biology: The Science of the Living World, Ernst Mayr. Belknap Press, 1998.
What is This Thing Called Science, Alan Chalmers. Oxford University Press, 1999.

THE BIG PICTURE

Earth's Restless Surface, Deirdre Janson-Smith with Gordon Cressey (Scientific Advisor). The Natural History Museum, London, 1996.
Leonardo's Mountain of Clams and The Diet of Worms, Stephen Jay Gould. Jonathon Cape,1998.
Life: An Unauthorised Biography, Richard Fortey. Flamingo, 1998.
The Dorling Kindersley Nature Encyclopedia. D.K., 1998.

THE COLLECTORS

Bright Paradise: Victorian Scientific Travellers, Peter Raby. Princeton University Press, 1998.
Images from Nature: Drawings and Paintings from the Library of The Natural History Museum. The Natural History Museum, London, 1998.

THE PLANET PROTECTORS AND THE TIME DETECTIVES

From the Beginning, Katie Edwards and Brian Rosen. The Natural History Museum, London, 2000.
Human Evolution: An Illustrated Guide, Peter Andrews, Chris Stringer and Maurice Wilson (Illustrator), Cambridge University Press, 1990.

TYING IT ALL TOGETHER

Discovering Dinosaurs in the American Museum of Natural History, Mark A. Norell and Eugene Gaffney (Contributor). A.A. Knopf, 1995.
The Natural History Museum Book of Dinosaurs, Tim Gardom with Angela Milner (Scientific Advisor). Carlton Books, 1993.

MAGAZINES

The New Scientist - http://www.newscientist.com/
National Geographic - http://www.nationalgeographic.com/

Web sites

Australian Museum online http://www.austmus.gov.au/
Collections and activities cover vertebrate and invertebrate zoology; palaeontology, mineralogy and anthropology. There are sections devoted to teachers and educators, and children.
Entomological Society of America http://www.entsoc.org/
World's largest organization devoted to entomology and related disciplines.
International Plant Genetic Resources Institute
http://www.cgiar.org/ipgri/
Extensive information and links to educational resources.
Internet Directory for Botany
http://www.helsinki.fi/kmus/botmenu/
Internet gateway to botany, giving access to more than 4000 resources in the subject.
Museum National d'Histoire Naturelle http://www.mnhn.fr/
Collections, exhibitions, research projects, publications and teaching activities. Many pages are translated to English.
National Museum of Natural History (Smithsonian Institution) http://www.si.edu/
Information on exhibitions, research projects and other activities. A selection of online exhibitions, with animated graphics and some interactive educational exercises.
The Natural History Museum, London http://www.nhm.ac.uk/
Main departments (botany, entomology, mineralogy, palaeontolgy, and zoology) provide details of research activities, collections and databases. Educational activities; exhibitions information and general information such as history and membership.
British Geological Survey http://www.bgs.ac.uk/
Centre for earth science information with links to all worldwide geological surveys.
Zoological Record Internet resource guide for zoology
http://www.york.biosis.org/zrdocs/zooinfo/
Index offers browsing by subject or by hierarchical arrangement of animal groups, using scientific names and a simple search facility.

NB. Web site addresses are subject to change.

Picture Credits

Unless listed below, all photographs are copyright The Natural History Museum, London. **Contents Page** (r) BAL/ DH; **p.4** (t) Paul Taçon, (b) BAL/ Ashmolean Museum, Oxford; **p.5** (l) BAL/ Museo Archeologico di Vill Guilia, Rome; **p.6** (t) BAL/ PC, (b) MEPL; **p.8** BAL/ Bibliotheque Nationale, Paris; **p.11** BAL/ Biblioteca Maciana, Venice; **p.13** BAL/ PC; **p.15** BAL/British Library/London; **p.16** (l) The Royal Collection © 2000 Her Majesty Queen Elizabeth II, (r) Graphische Sammlung Albertina, Vienna; **p.18** BAL/ Biblioteca Marucelliana, Florence; **p.19** SM/S&SPL; **p.20** BAL/ PC; **p.21** BAL/ Louvre, Paris; **p.22** SM/S&SPL; **p.23** (b) BAL/ PC; **p.24** BAL/ National Trust, Petworth House, Sussex; **p.29** John Taylor and Emily Glover (NHM); **p.30** Caroline Jones; **p.35** (tl) BAL/Linnean Society, London; **p.40** MEPL; **p.53** BAL/ DH; **p.65** Universal (Courtesy the Kobal Collection); **p.67** (r) Kelly Jackson and Iain Sime; **p.71** (l) Tony Stone Images/Alan Kehr; **p.71** (r) Sandy Knapp; **p.72** Southampton Oceanography Centre; **p.74** Kelly Jackson and Iain Sime; **p.75** SPL/ Ken Edward; **p.77** Sandy Knapp; **p.81** (l) R. J. Clarke, University of Witwatersrand, South Africa, (r) MBG; **p.83** NP; **p.84** PWD; **p.85** (l) SPL/ Karen Marks, NOAA Geosciences Laboratory, (r) SPL/ David Parker; **p.86** SPL/ Haxby, Lamont-Doherty Earth Observatory; **p.87** (l) SPL/ Professor Ian T. Millar, (r) SPL/ Science Source; **p.88** BAL/Ashmolean Museum, Oxford; **p.91** R. P.Hoblitt, USGS; **p.92** PWD; **p.93** PWD; **p.96** (t&b) LC; **p.97** (t) NASA; **p.98** (t) LC, (b) Emily Hedges; **p.99** (t) SPL/ Graham Ewens; **p.100** (b) SPL/Bob Edwards; **p.101** (t) Barbara Cressey/University of Southampton (b) BC/ Jeremy Grayson; **p.102** (tl, tr, bl & br) MP/AS; **p.103** (t) MP/AS, (b) SPL/ D. Phillips; **p.104** (tl & tr) MP/AS, (br) SPL/Kwangshin Kim; (bl) SPL/Dr Jopal Murti; **p.105** SPL/Mike Agliolo; **p.106** (t) NP; **p.107** (t) NP/ Derek Washington; **p.108** LC; **p.109** (t) BC/Stephen J. Krasemann, (b) BC./Gunter Ziesler; **p.110** SPL/ Astrid & Hanns-Frieder Michler; **p.111** NP; **p.112** (t) BC/ Orion Press, (b) BC/ Jeff Foot; **p.113** SPL/ Francis Leroy, Biocosmos; **p.115** James Mallet; **p.119** SPL/ Dr. Linda Stannard; **p.122** (t) Claudia Sprengel (Bremmen University) and Jeremy Young (NHM), (c) Lluisa Cros and José Fortuno (Institute of Marine Sciences, Barcelona), (b) Marcus Geison and Jeremy Young (NHM); **p.129** MEPL; **p.132** NP/ Hugh Vlark; **p.133** (b) Barbara Cressey/University of Southampton; **p.141** MBG/ Barry Hammel; **p.146** (b) Jan Fabre; **p.147** (t) MEPL; **p.153** Sandy Knapp; **p.155** Greenpeace/ Morgan; **p.158** BC/ Eckhart Pott; **p.160** Bryan Kneale; **p.166** BC/ Nicholas Devore; **p.167** LC; **p.169** (t) Joe Cann, Leeds University; **p.170** (t & b) NASA; **p.171** NASA; **p.173** Linda Pitkin; **p.177** Peter Andrews; **p.178** The Field Museum, Chicago. Neg no: GEO86127c; **p.180** (t) Brian Rosen; **p.188** (t & b) NP; **p.191** (r) O. Louis Mazzatenta/NGS Images Collection; **p.193** (t & b) *Science* Vol.282 No.5392 © 2000. American Association for the Advancement of Science; **p.196** Per Ahlberg/Nature (vol:385, p.489, 1997) Macmillan Magazines Ltd; **p.197** NASA; **p.198** Randy Korotev, Washington University; **p.199** NASA; **p.201** Linda Pitkin; **p.204** SPL/NASA; **p.205** NASA; **p.206** NASA; **p.207** PWD
AS – Andrew Syred, BAL – The Bridgeman Art Library, BC – Bruce Coleman Ltd., DH – Down House, Downe, Kent, LC – Laurie Campbell, MBG – Missouri Botanical Gardens, MEPL – Mary Evans Picture Library, MP – Microscopix Photolibrary, NP – Nature Photographers, PC – Private Collection, PWD – Perks Willis Design, SM – Science Museum, SPL – Science Photo Library, S&SPL – Science and Society Picture Library

Index

G

H

I

J

K

L

M

N